LEAN AND AGILE

SERVICES

I0479389

Inspiring Service Industry

Change Through Story:

The Lean Thinking Way.

Dr Chukwuma Destiny Ogbonnaya

Dedication

This book is dedicated to the City of Manchester, United Kingdom, where I experienced excellent services that significantly shaped my thinking and experiences.

Table of Contents

Preface

I write books to document experiences in my life that I believe can be valuable to readers if shared. I have read several diverse books and listened to audiobooks, and I wish to contribute to that pool of knowledge that continuously improves human conditions on earth. My first book *"The Scholarship Handbook for Undergraduate, Masters, PhD and Post-doctoral levels within and outside Nigeria"* was written after trying so hard to get an overseas scholarship to no avail. I documented my experiences to guide others on how to explore local and overseas scholarship opportunities. Writing the book was somewhat therapeutic because I was healed of the disappointments of not getting a scholarship immediately. I moved on after writing the book. The book gave me a sense of achievement, after all, since I had something tangible to share with others. It proved very useful to many scholarship seekers based on the feedbacks I received. Surprisingly, I later won fully funded overseas Masters and PhD scholarships to the UK.

My second book *"Letting go"* was a fictional book that I wrote on love, courtship and relationships after my experiences of my courtship, hearing other people's story and reading books on relationships. That the prove of love is, sometimes, in "letting go" is counterintuitive because people are more inclined to hold tight to love. The book was written as a screenplay which made it highly visual.

My third book *"Personal Financial Management using Employment Life Cycle Fund (ELF) model"* was written after my 5+ years of working as a consumer banker in a commercial bank. In addition to being a banker, I embarked on the most ambitious venture I have ever taken in my entire life. Coming from an engineering background, I set out to become a chartered accountant in 2 years. This meant that I must not fail any of the papers at all the stages of the exam. I progressed to the final stage in 1.5 years before I paused the exams to focus on my Masters. The book provides strategies for effective personal finance for employees based on an approach conceptualised as employment life cycle fund model.

My books *"Thinking, Knowing, Doing and Being"* alongside *"The Questelligence Workbook: A companion to Thinking, Knowing, Doing and Being"* were written after my PhD programme. I was very productive during my PhD with 10 publications in international journals, 2 conference and 4 seminar presentations. The book documented some topics that I had the opportunity to critically think upon and reflect on during my PhD. They were not purely academic topics *per se* - the contents were philosophical, and a critical thinking on random but contemporary topics. The book also contains an important theoretical proposition on the questelligence theory and theory of thinking.

This current book "LEAN AND AGILE SERVICES - Inspiring Service Industry Change Through Story: The Lean Thinking Way" was inspired by my experience during a surgery in a world-class hospital. It is presented from the perspectives of a patient and a teacher of lean thinking at the university. I reckon that my experience requires a documentation in a manner that can be valuable to service-based organisations. In doing so, I have also reflected on other services that I have used or delivered to enrich the book. I hope that this book will add a quantifiable value to you and your organisation now or in the future. I believe that lean thinking can contribute to continuous improvements in service industries.

Dr Chukwuma Destiny Ogbonnaya
Author
Chukwumao@oef.org.ng

Introduction

Lean thinking methodology is about getting more value from a system with less resources which is an accessible description of the concept of efficiency. Efficiency is how much useful outputs that a system can produce based on the inputs it receives. Both manufacturing and service systems receive diverse material and non-material inputs and transform them into tangible, intangible or a mixture of tangible and intangible outputs. A portion of the outputs are valuable while the balance is not valuable. The concept of lean thinking is about reducing the portion of the output that is not valuable which implicitly increases the portion of the output that is valuable.

As an illustration, if the total monetary value of generating a profit of £100,000 from a service system is £1,000,000. The efficiency of the service system is 10%. This means that £900,000 or 90% was still within the system. I agree that not all the utilisations of the 90% of the input would amount to wastes, but how do we know? Lean thinking seeks to address the question of how we can know. By systematic applications of lean thinking, lean tools and techniques, all the non-value adding utilisations of scarce resources can be removed. Supposing that the 90% waste was reduced to 70%, the new balance with the same input would be £300,000 of profit and £700,000 of waste. The principle of continuous improvement implies that the £700,000 could be reduced further to £600,000 and below, until an optimal point is reached where nearly all non-value-adding activities are eliminated and only value-adding activities can remain to delight the customer.

The major distinguishing feature of a product and a service is the tangibility of the value that customers consume from the system. Whereas the customer ultimately pays for value in all settings, the form of value in service industries is often intangible and experiential. Thus, culture, backgrounds, emotions and feelings, *et cetera*, which vary from one customer to another, may influence how a customer estimates the significance of a value created by a service organisation. For a tangible product, its functionality, quality, reliability, cost and delivery may be considered by the customers.

Although this book used a storytelling approach for presentation to make it accessible to a wide range of readers, the academic methodology is based on action research. I am acting as the patient or customer and making interpretations based on my knowledge of lean thinking. Passing through a service system as an informed lean thinker allowed me to acquire information that can be useful for other service organisations. I teach lean operations, agile manufacturing, and supply chain management in the university. An uninformed patient may pass through the same service system without recognising the lean principles or interpreting them in a manner that they can be useful in improving other service organisations. I invite readers to front load their experiences in diverse service industries so that they can reflect on their experiences based on my propositions and analyses. As an academic, I would like to get an email on any aspect of this book so that we can continue the conversation. Of course, our experiences may likely differ or appear similar in some cases, but the important bit is to reflect on your experiences based on the lean and agile principles.

Lean manufacturing principles have long been applied to service industries to eliminate wastes, streamline operations, improve customer experience, and increase profitability. Since "wastes" will be used multiple times in this book, the 7 wastes in manufacturing, as defined by Taiichi Ohno, are defects, overproduction, waiting, transportation, inventories, motion, and excess-processing. His argument was that these wastes do not add value and should be eliminated from a manufacturing system. Unutilised talent was later added as the 8th waste in manufacturing. Value, in this context, implies any process activity or feature of a product or service that a customer will be willing to pay for.

Not all lean manufacturing principles may be applicable in all service industries to the degree they have been applied in manufacturing. Nonetheless, the scope of lean thinking needed for continuous improvements in service industries is just as much as will be sufficient to create more value with less resources from service systems. It may well be that the manufacturing sector may benefit from applications of lean thinking in the service industry in the future. This is because value creation is intrinsic to lean thinking and it can be created by a manufacturing or a service system.

This is not a pure academic publication even though academic principles underpin most of the discussions and analysis. The names of service organisations were not mentioned to avoid appearing to be promoting a particular product or organisation. However, cities and names of countries were mentioned in some of the stories.

The overall strategy for presenting this book is to make it accessible to varieties of readers including academics, managers, staff, students, lean thinking practitioners, policymakers, etc. The writing style is also a mixture of storytelling, academic writing and reflective practice so that my experiences in the hospital can be linked to the theories and practices in lean thinking and business management.

Furthermore, thematic approach instead of structuring the content into chapters is adopted. This helps to show readers how the themes under discussion can be applied in other service-based organisations and industries. Principles are general in their effects, but they are contextualisable. As such, principles that satisfy customers of a manufacturing firm may also satisfy customers of a service firm if the principles are properly contextualised and adapted. This implies that, whilst I narrate my experiences, readers can be reflecting on their experiences of services they have received and service organisations they have received service from or worked for. Readers can also research further the themes that featured in this book as they also represent contemporary topics in worldclass service and manufacturing organisations.

Services that can creatively apply the principles advanced in this book may be a private or public sector of any scale. Some the services that can benefit from lean and agile thinking include but not limited to government departments or ministries or agencies, event management services, utility services, NGOs/charities, salons, hospitals, financial services, law enforcement, legal services, distribution services, hospitality and tourism, transportation, educational services, sports, entertainment, etc. Where relevant and appropriate, any of these sectors can be used to illustrate the theme under discussion to buttress my hospital experience. The themes are not arranged in any order and some themes may overlap. Therefore, the thematic approach is the most appropriate and beneficial in the current context instead of attempting to narrate the story chronologically.

Theme 1: Voice of the customer

The British culture of respect and courtesy, and the principles of person-centred care in healthcare services delivery influenced my overall experience in the hospital. In service delivery, the value of the service delivered by an organisation is valuated from the perspective of the customers. Regardless of the amount an organisation spends to deliver services, if customers do not value it, it is a waste of resources from the perspective of lean thinking.

When I got to the hospital, I answered the questions from the triage nurse that attended to me. That was when my voice as a customer (or patient to be contextual) was captured. I had the opportunity to describe, as much as I could, what brought me to the hospital. Notably, it is a need that usually compel customers to approach any organisation. When I met the 1st doctor (a General Practitioner (GP)) and subsequently the 2nd doctor (a surgeon), they still asked questions on what brought me to the hospital. The GP took detailed digital notes to the extent that I wondered why the surgeon wanted to hear the voice of the patient again. Doesn't she have access to my records, I wondered? But, I was patient to narrate my case because I am aware of the importance of the voice of the customer. Each practitioner sought to establish an understanding of my needs by hearing from me directly.

The voice of a customer is beyond what a customer is saying because words are not always enough to convey needs comprehensively. A direct observation of customers and their non-verbal cues may provide intelligence on the actual needs of the customer. Yet, the starting point is what the customers can communicate as their needs. The needs of customers should not be based on assumptions but on the customers voices.

The voice of the customer carries the imprint, call it the DNA if you are a fan of biology, of the needs and expectations of the customer. Ignoring it may lead to delivering below their expectations. It can cause the organisation to solve the wrong problem for the customer. This will be wasteful! A misunderstanding of the needs of the customer may even lead to a defective service system design. Meanwhile, the system design and configuration determine the behaviours in the system and how the system interacts with its environment. At each crucial point in the transformation process characterising a service system, the voice of the customer should be listened to, where possible. For instance, I gave consent to be operated on. I choose my breakfast, lunch, and dinner. Exactly what I ordered was served and it gave me a kind of satisfaction because I had no reason to complain even if I did not like the food.

If your customers are complaining, can you please check if you are listening to them. I think that some organisations are so noisy and busy with non-value-adding activities that they rarely hear the voices of their customers who are yearning to be heard.

As a teacher in the university, I take the emails, feedbacks, and questions of my students seriously. Those for me carry the voices of my customer (or my students). My first task is usually to create an atmosphere where all sorts of questions and feedbacks can come through to me unhindered. I say things like "no question is a nonsense question because your question can inspire orders to think". My email is available to all the students. I respond to emails as swiftly as possible, and I take additional actions when required. I brought my customer service culture from my banking background into my teaching, and I am certainly enjoying it. Transferring the principle of customer service from my banking background to higher education teaching was an example of what I alluded to earlier as principles being adaptable and transferrable. I look at each student as my customer and I do my very best to meet their needs. So, for me, the voices of my students are fundamental in meeting their learning needs. This should be true for customers in general as I cannot conceive a service system where the customers do not wish to be listened to.

In one of my modules which required the students to work as a team to produce a conceptual design, detailed engineering designs, manufacture, test and present their project, I used questions from the students to create Frequently Asked Questions (FAQs). It is validating to demonstrate to students (or customers) that their questions and feedbacks (or voices) are valuable. For me, it is a means of reinforcing curiosity and questelligence in students. Adopting an action research disposition, I use the opportunity of responding to questions to guide the students to achieve the assessment criteria as well as explain the engineering principles and professional skills that they would acquire if they achieved those criteria. Education sector is a service-based sector, and the student will leave the system with whatever they are able to learn. In other words, schools transform students and their experiences based on the overall curriculum, non-curriculum and extracurricular activities happening in the school as a service system. As a teacher, I have one goal - to facilitate an environment where students will develop the skill sets, knowledge and confidence to solve real world problems.

So, in service organisations, the voice of a customer is fundamental and should be captured in detail and considered carefully before, during and after the service delivery. This will enable the organisation to implement continuous improvements that are directed towards meeting the current, latent and future valuable needs of the customers.

In manufacturing, the Kanban system works in a fashion that materials and control information are pulled by the customer instead of being pushed to the customer by the manufacturing system. In service delivery, ideally, services delivered to the customer needs to be pulled as well instead of being pushed to the customer. That is, service operations should respond with series of activities to meet the request of a customer. Since services deal with intangibles in most cases, the voice of the customer needs to be heard clearly so that the mechanisms to respond adequately can be activated appropriately.

Theme 2: Communication

Communication is broader and more complex than the voice of the customer. Communication could be among staff, between a staff and a customer, between the organisation and external organisations such as their suppliers, government, unions, and other stakeholders. The focus here will be limited to my experience of communication in the hospital and in other service industries.

In the hospital, there were abundance of multimedia communications. At no point was I confused about what to do but there appeared to be long durations of waiting. Waiting periods in services may require effective communication to keep the customer updated. In lean methodology, waiting is a form of waste. In services, waiting manifests in different forms including long queues and untreated customer requests. Because customers are part of the transformation processes in service delivery, the waiting of customers should be considered as a waste. Customers should not be kept waiting longer than necessary. When customers wait much longer than they need to, it may indicate that the service design is defective or that the operation is inefficient. The root cause analysis (Ishikawa diagram) can be used to understand why customers are waiting for long. The 5 whys promoted by Sakichi Toyoda can be used to get to the root cause. I am developing a questelligence framework for systems thinking based on seven domains of questelligence (objective, people, time, place, process, reason, specific). It is an approach that could promote interpersonal communications, creation and analysis of systems.

Communication is a crucial tool in service delivery. Effective communication and guidance of customers can enable them not to make mistakes which will end up taking the time of the staff to fix. Effective communication can enable customers to understand the service processes and their co-creative roles in the service system. Effective communication can nudge positive behaviours from the stakeholders. At times, all that the customers want is an explanation of why they are waiting and how long they need to wait. Being frank with customers is better than giving false hopes that may dash their expectations. In train stations, the announcements are beautifully communicated that I enjoy listening to them. I learned to close communications with "I am happy to" address further the concerns of customers at a train station in Coventry. When I heard the announcement on my way to Birmingham, I pondered on the difference the addition of "happy" made. I wondered if some customer service officers attend to customers, but not happily. Stating that they were happy to help customers is an open invitation and assurance to all passengers to ask questions or make enquiries. Such a service environment is empowering to the customers.

A service operation needs to be designed to provide control information that can impact positively on the experience of the customer. A customer may experience a defective service. The negative experience can be managed with sufficient communication to explain what went wrong to the customer in addition to taking appropriate actions to remedy the situation. Train stations or airports have not stopped making announcements to passengers when their services are late or cancelled. Imagine that there are no such communications at the train station or airports. It will be troubles everywhere.

On the other hand, a customer may have a good service delivery with no major incidence but with insufficient communication. The customer with an incidence that were managed through effective communication and appropriate actions may rate the service higher than the later customer who has no encounter. Encounters (positive and negative) create memories. However, the memories must be positive. This does not mean that the customers should be disturbed unnecessarily. Yet, it may imply that moments of positive encounters can be designed into the service processes. Think about services that you have received in different service sectors. Those companies that left you with positive memories are likely to win your loyalty, but the memories need to be anchored on a satisfactory experience. The communication processes should aim to identify incidents that can create negative experience in the customer before it gets to the point that customers are outrightly angry. Some services collect feedback immediately after delivering a service to a customer. The timely analysis of the feedback data may indicate what customers are feeling and thinking about the service organisation.

There should be a mechanism within the organisation to effect continuous improvements based on the insights from that feedback data and other relevant internal and external considerations.

Digitalisation and digitisation of services reduce direct physical communication from staff because there are channels for communicating with the customer. Digitalisation will be discussed in detail under Theme 23.

Communicating with customers may contribute to reduce their risk exposures. Recently, my bank integrated credit score checking service to the banking app. Of course, in credit-based economies, credit score is very important as it measures the credit worthiness of a person based on how they have treated past and current protected and unprotected credits. I got an email alert that my credit score dropped by 55 points. I investigated the reason through the app. It turned out that a water utility company opened an unauthorised account with my name and recorded that I missed two payments. When I called the water utility company, the customer service officer apologised for the incidence and committed to close the account as well as work with their partners to reverse the 55 points. By informing me that my credit rating was degrading even when I had no credit with the bank, I was delighted.

Communication can be verbal or written; it can be in digital or non-digital form. Depending on the nature of the service, the senses of the customers can be engaged through design and careful delivery. Some service organisations place their mission and vision on the walls. Customers read them and at times attempt to hold staff accountable to the company core values. In my book "Thinking, Knowing, Doing and Being ", I stated that communication is the most important ingredient in human interaction.

Misunderstanding, communication breakdown, ineffective communication, or absence of it in service systems create wastes. Such wastes manifest as waiting, defective service delivery, transportation, inventories, and motion wastes.

Inefficiencies due to ineffective communication are resolvable by providing more and timely communication. However, internally, addressing failures in service delivery due to communication breakdown should focus on the flow of control information. All parties involved in meeting the needs of a customer should be aware of the request through an effective channel of communication.

Theme 3: Engaging the five senses of customers

The notion that the customers may be the actual "object" that is being transformed in some service systems makes it important to design the service system to engage the senses of the customer. The five external sense organs that receive stimulus from the environment are the eyes, nose, ear, skin and mouth. The brain is the sixth sense organ but it is an internal sense organ and it processes consciousness. Consciousness is very important to services because services deal with experience, feelings, emotions, intangible values, and minimal tangible values. I know some readers may question why I have included the brain as a sense organ. I have already advanced an argument that consciousness which is the generated in the brain should be considered as a sense in science in my book "Thinking, Knowing, Doing and Being ". I will proceed with how the argument of consciousness as a sense applies to engaging the sense of consciousness of customers (or awareness) in a service environment.

To underscore the importance of consciousness, suppose that you are in a service and an object dropped suddenly on the floor and scared everyone. When people realised that it was an air freshener container, they returned to the queue and the service continued. Now, it is likely that the sound and the sight of the object contributed to the scare but what of the consciousness of personal safety? Again, would the computers or chairs in the office be part of the things that would be scared? The answer is no; and the explanation is that they will not be scared because they are not conscious. The responses of customers to events that may have been received through the physical sense organs which will culminate at the consciousness of the customer within his/her context or environment. Suppose again that you hear that an organisation's IT infrastructure was hacked, and all the data were breached and available on the internet. Will the way you respond to the news be the same if you were a customer be the same if you were not a customer? Certainly not. The consciousness of the fact that your data may be part of the data breached may likely stimulate your sense of data security and that consciousness may influence how you respond. If you are a customer, you may likely read more news, watch the TV or visit or contact the organisation to get more information. Regardless of how sophisticated artificial intelligence tools will get, the ultimate beneficiary of the services they deliver will be human beings with emotions, feelings, impressions, consciousness and culture. Therefore, consciousness should be considered as a sense that should be designed into a service system.

The experiences of the customer are connected to the interpretations they give to stimulus they receive through the sense organs, perhaps in comparison with their previous experiences or expectations. The experiences may be pleasant or unpleasant.

Immediately I entered the accident and emergency (A & E) area of the hospital, I did not find it difficult to use the visual directions, clearly positioned on the hallway, to navigate my way. Each door post or a transition to a new area has a label. The visual management, which is a lean technique, of the environment was outstanding. I remember when I went for a CT scan, I looked at a door and saw a health and safety warning on the door. The warning indicated that there should be no unauthorised entry into the room because of a risk of electromagnetic field. In fact, I moved three seats farther away from the door after reading the caution. Visual management was powerfully used for different purposes in the hospital including for health and safety warnings, directions, locations and guidelines.

Visual management can be used where seeing could communicate an information better. In service industry, visual management can be used to communicate health and safety issues, locations, directions, performance, targets, plans, etc.

Given that service industries deal with experience engineering, it is strategic to leave a lasting impression on the mind of customers when they visit the service location. Also, the visuals should be creative and rich enough to leave a lasting impression on the mind of the customer. I remember one secondary school that I visited some time ago. As I climbed the stairs, I saw periodic table and some scientific stories and facts painted beautifully on the walls. When I finally arrive at the office of my key contact person, I saw images of diverse people which suggested to me that my host embraces diversity and inclusion because one could see all races represented on the pictures. Inductively, I left with a generalisation that the school may well be promoting diversity and inclusion; and this was corroborated by the diversity in their student cohort that I taught in a widening participation programme. Even in the hospital, you could see images of people from different race and that was a powerful message of a work environment that accommodates diversity.

Depending on the service, other senses can be engaged. Take for instance the sense of feel through the skin. In Africa, the temperature is usually so high. It makes people so uncomfortable that people pay premium to travel in air-conditioned buses. I can recall vividly how it feels to walk into an air-conditioned room. Walking into a banking hall in Nigeria is like securing a refuge from the hot sun. The banking halls are usually airconditioned and some banks use air fresheners to engage the sense of smell. Radio is used to engage the sense of hearing in bulk cash area where people may spend a reasonable length of time. Overall, the idea is to target as much senses as possible depending on the service process design.

As digitalisation deepens in service sector, the user interfaces of websites and mobile apps need to enrich the customer experience. Virtual places or online platforms have become as real as physical places because most services are now delivered virtually. Therefore, service organisations need to engage professionals to design their virtual/online platforms to provide customers with excellent experience. Even on a virtual environment, the senses can still be engaged particularly the audio and visual senses. The audio clips, images, videos and colour combinations and the general graphics design contribute to the experience of the customer.

Theme 4: Being and appearing organised

Given that 5S (Sort, Set in order, Shine, Standardise, Sustain) was popularised in manufacturing of products in Japan, one may wonder if it is applicable in service delivery. I looked out diligently for where 5S was applied in the hospital. 5S is a technique for workplace organisation to facilitate efficiency, effectiveness, safety, and communication.

As I passed through the hospital, I saw how things were organised and cleaned. I saw nothing that I felt may pose a health and safety risk. Everything was pretty in their place and everywhere looked tidy and organised. I reckon that a hospital environment should ordinarily be clean to control diseases. However, such expectations should not be taken for granted when they are met. I kept looking for more applications of the 5S pillars. A disordered work environment wastes time and causes stresses. Think of looking for a form on a desktop that is not organised using folder. More time will be wasted compared to when the files are organised using folders. The desktop can be likened to the physical display space of a service organisation. Applying sorting on the user interface design may imply that the items such as the links, forms, portals, etc should be organised, classified and sustained in a manner that customers can self-navigate the app or website with ease to carry out their transactions on their own.

While I was at the hospital, at some point, I asked for a pair of socks, a bed sheet, a pillowcase, a new hospital dress and a towel. These items were brought to me in less than 5 minutes. As a lean thinker, without following the staff into the store, I imagined that they must have implemented 5S in the store that enabled the staff to respond so fast. I reflected on their responsiveness based on a black box model. Indeed, in some service industries, the customer may never know how organisations configure their operations to meet their needs, but they can feel and share experiences of how their needs were met. So, based on a black box model, I imagined that things were clearly sorted and arranged to increase agility.

Items can be organised using containers, shelves, or storages of some sort. Towel, socks, bed sheets, pillowcase, duvet are all textile materials and can be kept in the same store. Logistically, each ward appears to have a nearby store. This will save the staff the time of walking to a central store that may be far away. This will be transportation waste.

I thought about the alternative situation in which all the textile materials are stored together such that it will require sorting to find each type of textile material whenever a patient makes a request. This may be contrary to common sense, but lack of organisation can waste the time of the staff, induce stress, and leave the customer dissatisfied. Being organised is fundamental to agility because of the increase in visibility of items and processes.

P:D ratio inherently measures how responsive an organisation can be based on their service design. P is the total time it will take an organisation to process and meet the request of a customer; while D is the time the customer needs to wait from the time the order is initiated to the time the order is fulfilled. The P:D ratio of some service organisations is expected to be low compared to manufactured goods. Manufacturing uses postponements and customer order decoupling points to reduce the P:D ratio. Systems that deliver services are configured to be agile enough to respond to the needs of the customer. Think of the transport or taxi services where customers make an order and then wait for a taxi. If a customer patronises two taxi companies, the company that may likely win the customer's loyalty is the one that is configured to meet valuable needs of the customers. When I use taxi services, including those that I call and those that I initiate a service request using an App, I expect that the P:D ratio should be small. I expect that it should be cost effective. The car should be in good condition. The driver should be professional. Sometimes reliability is more important to me than cost - I simply need a taxi at whatever cost. These are expectations of a typical taxi customer. Operators of a taxi company needs to configure the service to pursue certain business strategy and performance objectives. Nigel Slack et al has listed the five performance objectives of manufacturing and service operations to include cost, quality, dependability, speed and flexibility.

In an office environment, the use of filing systems, cabinets, folders, colour coding, etc to implement the principles of 5S may reflect on the productivity of the staff. Some triggers of stress may be as simple as being frustrated about looking for a piece of information or document. This is avoidable by implementing the 5S pillars.

Workplace organisation should be implemented at the organisational levels. The 5S pillars is not just what individuals should maintain to organise their workspace, it should be a culture. If the principles are applied across the organisation, team members can easily blend into new environments and roles since they are aware of how the organisation is organised physically, conceptually and culturally.

Theme 5: Poka Yoke (mistake proofing)

Service processes should be such that the greater percent of the users can use it without creating new problems. A service organisation should avoid spending much of its time for resolving problems instead of creating values. Mistake proofing is applicable to products, processes and procedures in a service environment.

While I was in the hospital, the nurse showed me a buzzer by my bedside and asked me to press it if I needed an attention. I pressed it a couple of times and I was attended to. Something happened to the buzzer, accidentally. I mistakenly laid on it and the plug detached from the wall. False alarm was activated, and the nurse came and plugged the buzzer to the wall. I exclaimed in my head, this is Poka Yoke! I thought that I damaged the buzzer. I took a closer look at the entire design of the buzzer, and it was such that any excessive force on the cable will detach it from the wall instead of damaging it. The second time it happened, I plugged it back and then implemented my own Poka Yoke with my understanding of the root cause. I removed the buzzer from the bed and attached it to the rail of the bed where I can easily reach it.

Under this context, the buzzer needs to have 100% availability so that the voice of patience who cannot move can be heard. Imagine a situation that the patient is in distress and the buzzer malfunctions. The nurses may be unaware, and the patient's wellbeing may be at risk. Interestingly, the alarm would be activated, whenever the plug is removed or the buzzer is pressed, which is a good design. This is a good example of design thinking which will be discussed in Theme 22.

Tools, devices and pieces of equipment that are used for service delivery may influence the experience and safety of the customers. Nonetheless, it may also influence the experience of the staff. In the case of the buzzer, false alarms would distract the nurses and waste staff time. This was the case for the false alarms before I removed the buzzer from the bed.

Another interesting observation of Poka Yoke was in medication administration procedure and during blood sample collection for tests. Before medication was given, the name, date of birth and address were confirmed. This confirmation was also done before blood samples were collected and CT scan was done. So, at each point, there was efforts to ensure that treatment is being offered to the right person. As of it, to prevent errors, there may be points of identification, authentication and authorisation in service processes.

Poka Yoke can be implemented in processes or procedures of service delivery. The process flow of service delivery can show how the customer will transit through the system and where risk of errors and mistakes are plausible. With detailed risk identification and analysis, Poka Yoke can be implemented to prevent the risks from materialising. Take for instance, to use some software as a service (SaaS) or applications, the customer must have inputted their card payment details before use. Also, if subscription is not active, the customer would not have access to the software or the customer may be allowed a limited access. This helps to stop customers from using a software or an application without paying for it. This will prevent income leakages, bad debts, and the use of resources to recover payments for services rendered.

As the saying goes, prevention is better than cure. Poka Yoke is saying the same thing in a different way. Lean thinking focuses on the prevention of any form of waste; but its purview also covers removing existing wastes from the system. Therefore, Poka Yoke is a preventive approach in lean thinking methodology that can be applied for product design, process control and service procedures.

Although risk management will be covered under Theme 15, risk analysis is a way of determining where Poka Yoke can be implemented in a service system. For product design, Poka Yoke can be used to prevent accidents during the use of a product or during the operation of a machine in a factory. Of course, Poka Yoke can be applied in the design of products used for service delivery but the emphasis here is on applying it in the design of processes for service delivery. In quality management, getting a task right the first time is the effective approach to service delivery. It saves cost, removes wastes and enriches customer experience.

Theme 6: Flow management

As I was entering the hospital on Wednesday morning, I was in pains, and I was apprehensive. When I was leaving on Saturday afternoon, the pains were gone. I was happy and I was convinced that my health problem was properly addressed. I entered the hospital as an input and left as an output, but what changed was my experience and feelings. This is what it means for the customer to be part of the transformation processes. Again, consciousness as a sense is very important because many services target intangible values. For instance, which external sense organs are used to receive the stimulus of internal pains (not pains on the skin), self-esteem, or knowledge? These are values that only consciousness can enable the customer to assess whether they have improved or not. Patients know whether medical intervention has ameliorated or aggravated their pains. A person with low self-esteem that used a coaching service can tell if their self-confidence improved after the process. A student can explain how their knowledge and skills improved after a learning experience or a programme of study at any level of education.

To conceptualise a value flow, there are two skills required. First is system thinking which enables a visualisation of how components of a system interact at an atomic level up to a global level as the system transforms inputs into outputs. All products, regardless of their complexity was created by a manufacturing system, regardless of the scale of the manufacturing system. Similarly, for service delivery, there must be a service system regardless of the simplicity or the complexity of the service system.

Systematic thinking is a methodical or procedural approach for creating or analysing the interactions or non-interactions of the components of systems to visualise the flow of value through the system. This enables a lean thinker to create and visualise the cognitive map of the process-time-space sequence and logic of how system resources interact to create value from the system. When I was leaving the hospital, I was full of thanks, I wanted to thank someone; but who should I thank in particular? In fact, I felt happy that I thanked each person that attended to me profusely as I passed through the processes because I knew that they were adding value to my wellbeing.

Value stream mapping, popularised by John Shook and Mike Rother, is usually conducted upstream and not downstream. That is, value is traced from when the order of the customer is fulfilled up to when the customer placed the order. To summarise some values added to me starting from my exit point, the pharmacist gave me medication based on doctor's prescription; the doctors did surgery based on the combined results of laboratory scientist and radiologist; the lab scientist and radiologist perform tests based on the surgeon's recommendations; the surgeon recommended tests based on the preliminary examinations of the general practitioner; the general practitioner conducted the examinations based on the initial documentation of the triage nurse; the triage nurse made documentation because I presented myself. Values were added at each of the stages.

Process mapping shows the steps taken to accomplish a task – it does not matter what the task might be. Process mapping is relevant to products and services because processes transform inputs into outputs within a manufacturing system or a service system. In a manufacturing system, the two key flows to track value are materials and control information flows. The control information are the actual information and communications that are required to transform input materials using machines, methods, man, and money.

In service systems, materials are not always involved as in manufacturing systems. Rather, control information and the customer flow through the service system. It is important to highlight that it is not just the customer but the experience of the customer that should be targeted for performance measurement. This can be seen when customers provide feedback. Happy and satisfied customers will say nice things about the organisation while disappointed and sad customers will complain, and some may even swear never to return.

How can the flow of a customer, customer experiences and control information within a system be managed? The answer lies in the type of the service system and how the organisation wishes to configure its operations. Whatever the design may be, value should flow smoothly, efficiently and continuously through the service system. Henry Ford was the inventor of flow production in manufacturing of cars. The principles of standardisation and flow production are now standard practices in all sectors of manufacturing. Implementing flow management will tilt towards agile services because the aim of flow management is to shrink the process time to the barest minimum. The faster a customer is served the faster a service organisation generates revenues. Again, the implementation of flow management in service industry depends on the nature and configuration of the service system.

Theme 7: Innovation and Technology

The hospital that I visited is a world-class hospital by all standards. So, the latest pieces of medical equipment were available for patients. One of the hi-tech pieces of equipment that I used was the computerised tomography (CT) scanner which uses X-rays and computer to create detailed images of the internal parts of the body. In fact, after the CT scan, I was triaged through a faster route to enable my surgery to take place on Wednesday night, although it finally took place on Thursday morning.

Lean thinking in service delivery does not equal cost cutting; it is closer to cost reduction. Cost is used to imply the financial value of wastes and rightful expenses in creating value. Of course, profit is simply revenue less cost. I imagine that the CT scanner must be very costly as a fixed asset. However, a hospital that uses such equipment will likely get most of the diagnosis right the first time, which will lead to appropriate and timely treatment. Thus, such technology is not a luxury but a necessity for quality healthcare service delivery. This does not mean that all hospitals can afford it, but the point here is that hospitals will need investment in diagnostic equipment as a means of confirming the voice of the customer or hearing more than the customer can alter in a bid to provide adequate treatment.

For service companies, investing in innovations and technology is critical to improving the experience of the customer. As Moore's law continue to take effect, the cost and mass of computing devices will continue to reduce while the computing power will continue to increase. So, exciting days are ahead regarding what man will achieve with innovations and technologies as civilisations evolve. As of it, service industries need to leverage the available technologies in every generation to improve service delivery.

Lean thinking is interested in how innovation and technology are deployed in service delivery. Deployment must be informed by understanding how it will meet the performance objectives. Technologies need not be deployed because others are deploying them. However, if an innovation is fundamental to a service sector such that it will ultimately become part of service delivery in such a sector, it is important to be among the early adopters. A typical example was the introduction of Automatic Teller Machine (ATM) in banking services. The rate of adoption was based on the technology savviness of each bank, but it is inconceivable that a commercial bank, nowadays, anywhere across the globe is not using ATM to deliver cash to its customer. Point of sale (POS) is also another innovation in banking that has enabled customers and services to exchange value without dealing with a physical cash.

Artificial intelligence (AI) is expanding the innovation space in predicting what customers may like or need. By learning from big data, algorithms can recommend services to customers. AI tools are tapping into the human core (e.g., intelligence) and providing experiences that oftentimes appear to suggest that AI tools know people more than themselves or that it is eavesdropping on them.

A forward-thinking service organisation should be working with its innovation circle or research and development (R & D) or continuous improvement department, whatever the name of the department may be called, to see how AI tools can enhance their service delivery. As long as human beings are served by a service system, the system is dealing with intelligent agents and artificial intelligence can help a service organisation to systematically know more about their customers and their niche using AI tools, algorithms and big data.

Theme 8: Standardisation and optimisation

Standardisation of workflows makes automation and applied robotics plausible manufacturing. In service delivery, automation based on digitalisation and digitisation processes could make a huge difference in a lean and agile service delivery.

What facilitates standardisation in a service-based industry is professionalism and specialisations. When I was called for blood sample collection, the nurse had everything (syringes, labels, packages, etc) prepared based on the recommended tests by the doctor. I watched her implement the procedure sequentially. How she searched for veins, how she pierced through my body, the pace at which the blood flew, the quantity per syringe, how she terminated the process and ensured that there was no bleeding.

Service standardisation can be a competitive advantage for service organisations that have multiple branches or outlets. The organisation can create a lean and agile service system and replicate the system and processes at its branches. This can also cause their loyal customers to look out for them everywhere they go.

One day, my family decided to board the train, travel from Manchester to Liverpool to have a lunch and then return. Nothing else! We had no idea of where to have the lunch, but we hoped that Liverpool is a big city and there must be a place we can have a buffet. Few minutes to the Liverpool Lime Street where the train terminated, we started searching for a restaurant. My eyes caught a restaurant which serves buffet. I had patronised the Manchester branch a couple of months ago with some friends. My experience of the Manchester outlet was pleasant, and I hypothesised that if the standard was the same, then it would be a good place for the family to have the lunch. We used Google map to navigate until we found the place. When we entered the restaurant, the environment was welcoming, like what they had in Manchester; perhaps better. Fast forward to the end of the meal, I was not disappointed and the single factor for winning my blind patronage was that I was aware that strong brands tend to standardise their processes and systems.

Operating in the same location does not translate to operating at the same standard. In other words, two services may be offering the same services at the same location but the patronage by customers may differ. Whereas one would be the preferred choice and market leader, the other one would be patronised when the first choice is not available. In fact, some competitors tend to stay close to market leaders and offer reasonably good services to the extent that customers would take the follower as equally good as the market leader. Nonetheless, standardised and optimised service delivery is a more scientific means to remain competitive.

I had experiences with two County Councils on school admissions of my children. One has a robust application process, clearly mapped processes with instructions to guide the parent or carer. The other one was OK, to be polite, but the application portal was suboptimal, and the responses was not timely. My experiences with the two admissions department were two worlds apart.

I have used Visa application services of few countries. My experiences were not the same. The difference appears to be in the optimisation of the processes to make it accessible and easy to follow. After using the application portal of one country, I was wondering if a government of a nation could not afford a decent website for processing travel, visa, passport documents and citizens' requests. I doubt if anyone will not be unhappy after using the website. I was unsure that I did the right thing after submitting a request due to the mapping of the processes.

A service organisation must first standardise its processes. Once the flow of transactions is working well and value is flowing seamlessly, the organisation can aim to optimise the system to remove wastes, reduce costs, improve quality and enrich customer experiences. In developing countries, most of the government websites are there to fulfil all righteousness. Requests and enquiries made there are never responded to, and many processes are expected to be manual. At times, I wonder what the civil servants of such nations are doing if they cannot reply to emails or simple enquiries. Standard replies can even be created based on the collation of frequently asked questions. There is value in responding to a customer's enquiries even if the needs are untenable. Perhaps must people working in public sector do not know that their citizens are their customers. However, in principle, the same way civil or public servants are paid with taxpayers money is the same way staff of corporations are paid with the revenues generated from the customers. In comparison, private sector organisations appear to be more efficient than public sector organisations.

E-government is lean but issues of corruption does not allow the use of technology in public sector service delivery in developing countries. If you automate government processes and remove human agents, corruptions would be significantly reduced. This does not dilute the bureaucracy in public sectors, it simple improves flow of value and by extension quality service delivery to citizens. Lean thinking through standardisation could improve bureaucracy processes whilst preserving its form.

Theme 9: Continuous improvement

Medical service is unique because it deals with life. Patients come with unique cases. The teams in the hospital work quite hard to care for thousands of people. I always pay tribute to everyone working in the healthcare environment for the great work they do to save other people's life and keep them healthy. I am not a medical professional, and this book is not about service designs in hospitals. The book uses my experience in the hospital as a vehicle to discuss contemporary issues associated with lean thinking in service processes, systems and industries.

Last year, when I visited the same hospital, I was not completely satisfied with the services that I received. When I realised that I was to visit the same hospital, I told my wife that she should not expect me till the following day. I was so calm, and I was ready to simply wait until I get treated. However, it was a sort of surprise when I started progressing quite faster when I was triaged for surgery.

I felt that there was an improvement of my earlier experience, but it may well depend on the route I took. There may be other factors which cannot allow me to be conclusive. In my first visit, it was on a Saturday, and I had to pass through A&E route. In that first visit, I arrived at 1 pm on a Saturday and I left at about 10 am on Sunday. In the recent visit, it was on a weekday, and I passed through Urgent Treatment route. This enabled me to see two doctors (a GP and a surgeon) in less than 2 hours.

In both cases, the bottleneck appeared to be after seeing the doctors. This was understandably so. After a consultation with a doctor, tests may be recommended including but not limited to fluid (blood, urine) analyses, CT scans, X-rays, etc. After the analyses, the results must be returned to the doctors to take a clinical decision. Based on my previous visits to hospitals, the patient may be discharged with or without medication, referred to a specialist department or progressed to the next level of treatment such as surgery, as was my case. Why I adjudged my recent experience to be an improvement of my earlier experience is because I was not contacted after referral until I recovered fully. Nonetheless, in that first visit, I think that having an expert opinion that nothing extraordinary was seen after the x-ray and blood analysis gave me the confidence to go home and recover. This, to me, was very valuable. When a nurse called me few weeks later, the case was closed after some questions.

Based on Pareto principles and theory of constraints, the processes for continuous improvement in the hospital are the processes between the recommended tests by the doctor and when the patients are discharged, referred, or progressed.

Indeed, these processes are at the heart of medical science. The collection of samples or data on the patient, analysis of the multiple data sources to build a body of evidence, interpretation of the evidence based on scientific hypothesis and theories of medicine, reaching a clinical conclusion and decision-making on the cause of action based on the results, experience, best practice, and intuition. These activities are done by multiple professionals.

In manufacturing, Lean-Six Sigma has been used to reduce defects, variability and wastes in operations. Service-based organisations that need to implement lean methodology needs to be strategic about it. When I was in the bank, there was a bank-wide cost-reduction programme that saved huge amounts of money for the bank and it changed the way staff thought about cost. This was not tagged a lean programme, but the programme focussed on eliminating all sorts of wastes from all conceivable processes, activities, inventories, contracts with third parties, based on rationality and common sense. In the same manner, there are different organisations that may be implementing approaches to effectively reduce waste without tagging it a lean programme. Notwithstanding, the importance of being clear that an improvement programme is a lean programme is that it allows the organisation to be strategic and systematic in implementing the principles of lean thinking using a range of lean tools and techniques that have been developed over the years.

The lean tools and techniques in themselves do not solve problems. For instance, to use the Plan-Do-Check-Act (PDCA) cycle, which was developed by Walter Shewhart but popularised by W. Edwards Deming, there is a need to map the corporate goals to the operational objectives and the key performance indicators (KPIs). Even with Six Sigma methodology that uses statistical analysis to reduce process variability, the results of descriptive or inferential statistics mean nothing if they could not be properly interpreted in the context of the organisational goals and performance objectives. Overall, continuous improvement is incremental but that was at the heart of Toyota Production System. Continuous improvement programmes need to be systemic and systematic, and it should focus on causing value to flow more efficiently through a service system.

Causing value to flow is the goal of manufacturing systems engineering or service systems engineering. The notion of engineering of a service system is that a service system can be designed and controlled to behave in a certain predictable way. This involves the application of mathematics, social and physical sciences to create, operate and maintain the service system.

I think that I should introduce a new concept described as "value stream destruction map". The term is a scientific term and I think that its explanation might interest you. In thermodynamics, there are two ways to measure system efficiency. The first approach is the energy efficiency which simply calculates the energy output as a ratio of energy input and express it as a percentage. Theoretically, energy can neither be created nor destroyed but it can be transformed from one form to another. Thus, the sum of energy inputs must be equal to the sum of energy outputs, but they can exist in different forms of energy. As an example, electrical energy will flow into a light bulb and produce light (required energy) and heat (waste or by-product). In lean thinking, this reflects the concept of value-added and non-value-added outputs.

The second way to measure efficiency of a system is by calculating the exergy efficiency. Exergy efficiency calculates the maximum theoretical useful output a system can produce as it interacts with its immediate environment. Here, exergy cannot be conserved like energy, but it can be transferred or destroyed. How it was destroyed can be accounted for through exergy analysis. Exergy efficiency, therefore, is a more realistic way of measuring systems efficiency because it accounts for wastes in the system as a function of destruction of value. Exergy analysis can pinpoint the components within a system with the highest exergy destruction rate so that such components can be improved as a matter of priority. Theoretically, if a low exergy efficiency component is improved, it tends to improve the overall exergy efficiency of the system. The notion of continuous improvement is an acknowledgement that the theoretical efficiency limit of a manufacturing or service system has not been reached and that reducing the value destruction rates in the components and processes could lead to an improvement in the values created by the system.

Why I am excited about systems thinking and systematic thinking is that unrelated systems can be mapped together to learn lessons. This is while I believe that lean thinking lessons from manufacturing can be applied in service industry. This is why I believe that an average creative reader of this book can apply the principles in their unique context. Systems and systematic thinking are the thinking underpinning bioinspired design and engineering and biomimetics. Engineers look at nature and design artificial systems that mimic or look like nature.

My proposition is that lean thinking should be treated as an exergy model and not as an energy model supposing that exergy is equivalent to "value". Only when lean thinking is modelled as exergy analysis will the wastes and the value destruction rates in departments or units within a manufacturing or service systems can be studied and improved using lean tools and techniques. Therefore, value stream destruction map is defined as a map that would show the value that entered into each department and the value it produced and how all the departments contributed to the overall value created by the manufacturing or service system. The Sankey diagram can provide inspiration on how to construct the proposed map. I must state that this is hypothetical and reflects one the moments that I am writing with an academic voice, which is exploratory. This proposition is novel, and it appears suitable for academic research or an investigation by an R & D of a manufacturing or service company.

Theme 10: Collaborating with partners to eliminate wastes

I had a breakfast, a lunch and a dinner at the hospital on Thursday and Friday; and a breakfast and a lunch on Saturday before I was discharged. It may appear contradictory to state that the food served was lean and nutritious. The food is described as lean because I had no reason to waste the food. Waste of food in public and at home may be linked to poor planning, lack of appetite of the eater and lack of standardisation of quantities people can consume.

There were few factors that made me not to waste the food. First, on Thursday morning, I was asked to choose what I will eat on Friday morning, afternoon and night. On Friday, I was asked to choose what I will eat on Saturday. Without staying longer than 3 days, the pattern was clear that order for food is collected from patients a day in advance. The quantity of the food is such that a patient can finish it. The drinks are customised and packaged to ensure that patients with an average appetite can finish the drinks. The packaging and delivery to patients do not cause spillage of liquids or dropping of debris. This reduces the probability of littering of the environment with packages and containers by patients, which will need a staff to clean up. Thus, the collaborative planning of patients' meal and a careful control of the quantity appear to reduce the waste of food whilst offering the patients nutritious meals.

It is interesting to note that although a hospital is offering medical services, the foods consumed by the patients have different supply chain networks. Thus, activities within a service organisation may involve other service or manufacturing organisations. For instance, medications, apart from food, have different supply chains.

I have taken flights several times, but I cannot remember the things I ate. In contrast, I do not think that I would forget the experience of the food I ate at the hospital anytime soon. Some flights are booked months in advance, and I have wondered why customers cannot co-create what they will eat during the journey. With big data, the meal preferences of customers moving from one location to another can be predicted. Many first-time travellers from Africa to Europe usually talk about eating something that they could not recognise. Food is one of the ways to create memories and service organisations that offer food to their customers may need to move towards co-creating the meal with the customers. The variations in the menu will still be constrained as the customer will select from what is available on the menu. In fact, the current menu can be used while customers can be asked through feedback about other meals they would like to eat in their next trips. Some people like food and it may well be that bit that may concretise their experience and win their loyalty to the company, all other things being equal.

Services that require physical products as inputs need reliable supply chain partners. Collaborative planning can help the service organisation to receive the right quantity of inventories from the suppliers based on the order of the customers. In manufacturing, Kanban system and Just-in-time ensures that the orders placed with the suppliers and production are based on customer's order.

Theme 11: Professionalism

I was impressed by the professionalism of all the hospital staff that I came across. I think that professionalism in service industries can win the confidence of customers. Customers have a way of weighing risks as they deal with professionals or service organisations. For products, customers may not necessarily see the engineers and workers that design and manufacture the products. In service delivery, the clients or customers may need to pass through the process with the professionals. For instance, in legal services the client may not know the laws, rules, legislation, and court procedures. The client will depend on the guidance of the lawyer.

The conduct of a professional determines, to some extent, the experience of the customer or client. When I was in the bank, a customer was not able to open an account because his lawyer could not find his Corporate Affairs Commission documents. I was wondering how that happened. How was it possible for a professional to handle sensitive documents of a client in a manner that they could not be found. Lean thinking may include things as simple as developing a standard procedure on how to handle customers' sensitive documents and transactions. Lawyers handle pieces of evidence that may be massive but those that have systematic processes are more efficient. This applies to other services that handle sensitive documents. In the university, incidences occur where exam scripts of students get missing. This has a severe consequence for the student because it may affect their progression. A well-designed procedure for handling examination documents, scripts or artefacts can leave audit trails that can help pinpoint where the process flow was derailed. Improvements can be implemented as well.

Each profession has codes of conduct, ethics, rules and best practices. Those codes of conducts are drawn from ethical principles. Professionalism is good for the organisation as it enables the staff to offer quality services. It is good for the customer because they are in good hands.

From the point that I knew that I would have a surgery, I was very attentive to the communications that I was having. I was convinced that those discussing with me were very professional and my heart was at peace with the process. Interactions with professionals in a service system could incline a customer to estimate the risks involved in the processes. This is the reason some customers change their mind in the middle of a service process.

When my wife was about to deliver my first son in Nigeria, I passed the night in the hospital with her. As I continued to interact with the doctor, at some point,

I whispered to my wife that we will leave the hospital first thing in the morning. The interactions with the doctor and the nurses did not give me sufficient confidence for my wife to continue to stay in that hospital. Around 6 am, I called another hospital where my daughter was delivered, and I transferred her there. My confidence in the new hospital was higher. I was in the second hospital when my daughter was delivered, and I was more confident in the doctor. I left my wife with the nurses and the doctor to make a cash withdrawal at the bank. I was at the bank when I was called; and I was informed that she has given birth to our son. Afterwards, that hospital became the family hospital and for our visitors that need medical services. Service contracts are won or lost based on the confidence of the customer. Confidence may be based on perception, appearance and not reality but it matters when customers take decisions. If their perception is that the probability of a loss is quite high, or that their needs might not be met by the service system, their confidence will dwindle, and they may retract from passing through the service processes. At this point, many customers consider the alternatives, if available or try to rationalise the risks and proceed.

It is necessary to accentuate the importance of confidence in a process which may be the basis for trust in an entire system. Imagine a game in which a player is given two pipes, A and B, of equal diameter and equal length. However, A is a straight pipe while B is a convoluted pipe. The player has an option of inserting a coin into A and win £100 if it passed through to the other end; or win £10,000 if inserted in B and it passed through to the other end. If the coin did not pass through to the other end in both cases, the player will get £0 or nothing. What do you think the player would do? I understand that people have different risk appetites, but it is likely that more players may insert the coin in A. The reason is that the convoluted geometry is likely to be perceived as a risk factor. This may weaken the confidence of a player on the possibility of the coin passing through. Thus, a lack of trust in the convoluted pipe. This is like what happens when a customer is considering two investment options with different level of perceived risks. It will take a professional to communicate the risk appropriately so that customers would not exaggerate it and quit.

Supposing, again, that the situation is reversed such that the player can win £100 if the coin is trapped in A or win £10,000 if the coin is trapped in B. If the coin passes through to the other end in both cases, the player will get £0 or nothing. Where do you think the player will insert the coin? I guess in B. This is relevant to insurance companies. The insurance policy should dissuade customers from taking unnecessary risks since most customers may well cause the risks to materialise frequently if there is no demotivation. The insurance policy, therefore, must be designed and scoped to give the insurance company the confidence to sell it to customers. Insurance businesses make profit because they can manage the risks.

Services are very volatile because customers can easily go to the competitor. In banking, customers may have more than two accounts, but one can tell which bank that has won the confidence of the customer by checking the bank that is handling higher proportion of the customer's banking needs. A customer may also choose different products from different banks based on the competitiveness of the product and service delivery. I have two mobile telephone numbers. One company is good at international calls while the other is good at data. Service organisation within the same industry often try to be unique or outstanding in some offerings. There is a bank in Nigeria that specialises in merger and acquisition, and this has enabled it to increase access to more customers. There is a perception of competitiveness and robustness by customers. This competition strategy is facilitated by professionals and processes that enable effective mergers and acquisitions. Again, lean thinking is not about cutting cost, it is about achieving optimal efficiency and effectiveness. Professionals are vital in many services unlike in manufacturing where factories can be fully automated with robots doing repetitive tasks.

Some have argued that unutilised talents should not be among the 8 wastes in manufacturing. In services, unutilised talent appears to rank high. Services are delivered through three critical modes - self-service, technology-driven service, and staff-service. However, when the first two modes fail, the last line of defence is the staff-service. Once upon a time, I completed a form to exit the provider of my gas & electric. When I called the customer service, the officer was so professional that at the end of the day, I stayed with the company. I did not hesitate to tell the officer that I was staying because of her level of professionalism. I had a similar experience when my Internet service had issues. I reported it and it was addressed but the problem persisted. I reported it the second time, it was addressed but the problem persisted. When I called again, I was frank with the customer service that the quality of the Internet was unacceptable. At this point, I had made up my mind to give the company one more chance for the sake of the good services they have provided in the past. This time, the customer service escalated the problem, and it was finally resolved; and I stayed. Becoming conscious that my internet provider has the capacity to resolve problems reinforced my loyalty. Dependability, as a measure of performance of an operation, has to do with the confidence customers have on the system to meet their needs at all time and when it matters.

Many service organisations hire staff that do not provide that last line of defence to meet customers need and retain the customers. Targeting referrals through excellent service delivery is a form of marketing but some organisations are oftentimes preoccupied with acquiring new customers instead of serving the existing ones. Therefore, training staff and improving their professional standing may well prove to be beneficial for the service organisation.

Theme 12: Service environment

In this hospital, the environment was very clean, and the thought of germs and viruses hiding around the corner was toned down by the consistent efforts of the organisation to clean and disinfected the environment and the toilets. Usually, my approach to my health is to prevent getting ill. Of course, when there is a concern, I will see health professionals. I do not like to see blood and people in pains. When I was in secondary school, this was the reason I chose engineering over medicine and surgery. However, this has changed over the years as I grew older.

Although the hospital environment makes me uncomfortable, I have realised that it is usually a moment to be grateful for the good health I enjoy. Many people do not know that health is wealth until they take ill and get to the hospital. Visiting the hospital reminds me of the frailty of the human body and the need to care for it. By the way, I wish all readers of this book good health and happiness. Good health has a link to service delivery. A sick person is likely to deliver defective services. During COVID-19, many services were badly affected by the exposure of their staff, and some closed to keep the staff and the customers safe. So, there is risks associated with services that must be delivered only at a physical location.

When a service requires a customer to appear in-person, the service environment needs to be designed to support physical services. The service environment should be attractive, inspiring and memorable. The artwork I saw on the wall on my way to the theatre was like a childlike art, so beautiful and creative. I saw shapes and paintings. I remember thinking to myself that I need to capture those beautiful shapes and hold them on my mind so that I can dream about them when I get anaesthetised.

Surprisingly, the moment I was informed that anaesthesia was about to begin, I cannot recall having any dream. Rather, I woke up on a bed in a ward, with a doctor by my side telling me that the surgery went fine. I replied, "thank you". I tried to remember if I dreamt or not, but the period of the surgery appeared deleted from my memory.

In service delivery, the memory of the environment is part of what the customers will take home with them. Even in services targeted at children, an immersive environment may leave a lasting impression on them. There was an aquarium my family visited. It was a mind-blowing experience as we saw diverse sea creatures. I cannot forget that experience. My daughter is still asking if it was not time to visit the aquarium again. I also recall a day that my son refused to leave a recreational park because of how engaging and imaginative it was. For services targeting children, the environment and activities need to be designed with children in mind even though their parents will pay for the services.

Environment feeds the six senses with stimulus. Whenever I visit one Nigerian bank, I usually see mirrors positioned at strategic points. When the bank started, it was assumed that it was only for rich people because of the kind of ambience and environment they maintained. The building is usually located at a strategic location in the city, regardless of how costly it is. The benefit of standardising office building is that it enables the company to optimise the design and cost and benefit from a well-negotiated contract with suppliers. This includes interior designs that reflect the company colours and branding.

During Christmas last year, my family visited my cousin's family in Bradford. Coincidentally, 26th December was her birthday. We decided that we will all go to a restaurant that offers buffet service. When we got there, we enjoyed the food. However, at some point, my son needed to use the toilet and I needed to accompany him. When we got there, there was water on the floor and the toilet seat was very old and dirty. It was like I was being punished to assist my son to use the toilet. That toilet spoiled my day as the experience was not pleasant at all. If I go to Bradford again, I will not go to that restaurant. I will try a different restaurant. For a restaurant, the toilet area is critical. It may be more important to some people than the meals. Some people go to such restaurants to socialise and have a meal. So, the objectives that should be pursued is not limited to the varieties of food on the menu. The environment where people have such meal is equally important. Ambience and environment where services are delivered to customers need to be welcoming, assuring, relaxing, safe and memorable. Some have used branding, professional cleaning services, eco-design, artwork, etc to achieve unique ambience and environment.

Theme 13: Lean and Agile Service

When I realised that I needed to go to the hospital, I told my wife not to expect me till the following day. Generally, patients expect agile services from hospitals, but the health professionals know that the right approach is the lean approach. Some patients would yell, cry, beg, cajole, and do all sorts of things to be treated at the speed of light. Each patient yearns for an attention so that they can be relieved.

The health professionals know that they are dealing with life of human beings. They must be systematic not to mix up patients' records, samples, diagnosis, medications, treatment, etc. This requires efficiency, the lowest margin of error and not speed. *Ab initio*, there is a mismatch between how patients expect medical professionals to respond and how the professionals need to respond. Understandably, there are situations where agility is required, particularly under accidents and emergency, and the triage system confer agility in the case of hospitals.

There are services that need a good balance between agility and leanness. Organisations need not think that there is a general formula to adopt but should pursue a unique pathway for implementing a lean and agile processes based on their product and service offerings. Lean thinking focuses on removing wastes so that the only thing that remains would be value that customers would be happy to pay for.

I reckon that many service organisations are yet to understand the importance of systematically incorporating lean thinking into their service design and processes. Indeed, chaos does not create patterns, but a beautiful pattern can be created from a chaotic situation. A typical lean tool for handling chaos in manufacturing is "heijunka" or levelling of operations. In practical terms, whenever a service system is highly efficient, there is a deliberate design behind it given that design determines physical behaviours.

There was a religious organisation that announced at the beginning of the year that they would close at a certain time. In this case, time management was a valuable aspect of the service that was intended to be improved. Members needed to work to earn a living and pay for their bills. The proceedings of the service were adjusted to enable members to meet their spiritual, social and personal needs. The new closing time was complied with for few months, and this led to an increase in the number of people that attend the religious gathering. In many community-based organisation, clubs or associations where members are not under any obligation to participate, there is a need to design the activities and processes to achieve the objectives of the organisation whilst taking into cognizance the needs of the members.

Services where people are free to join and exit at any time should design their processes to encourage participation. The incentives to participate should be tenable and convenient to the members.

Quality, cost effectiveness and excellent delivery should flow together through service processes. This means that the best quality service should be achieved at lowest cost, and these should not affect the speed and flexibility of the delivery of the service. Quality, Cost, Delivery (QCD) trilogy is popular in project management. Likewise in service design, the flow of value can be examined based on the QCD management framework. The relevant components of QCD as they apply to a service system can be determined by the organisation with due consideration to the voice of the customer and the overall business strategy and prevailing realities.

Overall, the service design will be determined by the nature of the service. Clearly, some services may be for-profit while others are not-for-profit organisations. Service-based organisations can be designed with both lean and agile principles to facilitate the achievement of the strategic objectives. Therefore, by seeking the optimal balance of lean-agile model, efficiency and responsiveness to changes will be balanced based on the performance objectives of the service organisation.

Theme 14: Entertainments and amusements

Life is beautiful because of diversities in the world. I had always thought that the world would be an extremely boring place if everyone on earth was the same in every way. In the hospital, I saw a woman who was sick, but she was even more sincerely concerned about others. She was very friendly and quite outspoken. She called the attention of a nurse and explained that she needed to be attended so that she could return home. The nurse took her name and asked her to hold on so that she could confirm her details. As soon as the nurse left, she asked a woman by her side, who was visibly in pains, if she should also ask the doctor to attend to her. I was somewhat amused because her request was still on transit, and she has not received a feedback, yet she was even much more concerned about another patient. That level of pure empathy and unfeigned kindness was moving to observe and dare say that such emotion to care for and nurse each other has preserved the human species over time.

At some point, she came and sat next to me and told me that she would not want to stay isolated and far away from others. Her friendliness made me use her phone charger to charge my phone. Later, another middle-aged woman brought an older sick woman to the waiting area. The woman was very funny and loud. Even though I was in pains, I cannot recount how many times she made me laugh uncontrollably. She was not disruptive, but she was very assertive and funny. I wondered if she was aware that she was that funny. On several occasions, I saw people in pains laughing because of the way she was interacting with the older sick woman who was sitting on a wheelchair. Every moment of their interactions made me laugh, I felt my energy level surged, and my strength was renewed to keep waiting. She was such a phenomenal character in a positive sense.

In service design, what customers do while they are waiting can influence their perception of their overall experience. If lean thinking has indeed classified waiting as a waste in manufacturing, then it should be a waste in services. The waiting of a customer as a part of the transformation process in services is a waste. Waiting of their transactions are oftentimes the focus of waiting waste in service industry. Time is intangible but people know when it is wasted. In manufacturing, customers can be busy with other things until their products are ready. Customers can as well walk into a supermarket to buy finished products. In services that requires the presence of the customer, their time and space is confined to the location of the service during service delivery. A hairstylist or barber service cannot style the hair of a customer at the office while the customer is at home.

So, how might an organisation provide value to customers while they wait for the processes of service delivery. Some banking halls or telecommunications outlets keep customers entertained with TV or radio or music while they wait for their turn or while they are passing through the offices. Ultimately, for in-person services, the waiting time must be reduced by design or through offering valuable activities. Offering valuable activities does not reduce time but it takes the mind of the customer away from the boring moment of waiting idly. I like reading books. While I was waiting in the hospital, I started looking for anything at all to read. The battery of my mobile phone was drained, and I appeared cut-off from the rest of the world.

With the cost of digital devices decreasing, service organisations with significant waiting time may have digital or E-book that customers can read while they are waiting. Such devices can even provide more information about the organisation and its services. Some customers may like games, puzzles, mazes, etc, just to while away time. The downside of this is that it may cause congestion as customers may stay back to be entertained after being served.

This is not to say that the problem of long waiting time can only be solved by providing entertainment or value for the waiting time. The staff strength is also a factor that determines how long customers will wait in service-based businesses. For instance, the number of hairstylists may affect the length of time customers spend in a saloon. The higher the number of staff the faster the service delivery, but that will increase the cost of labour. There may be a need to invest in resources to create additional capacity to reduce the waiting time. How about applying the principles of postponement so that a customer can choose the hairstyle and materials in advance to reduce the duration of the service. I think that weave on is ready to fit which is prepared and appear to fit into the definition of postponement.

A booking system that can enable customers to schedule an appointment before they arrive at the service point could control the number of customers at the service area. A typical example in the use of scheduling to reduce waiting time is in Visa processing. Customers arrive and exit through the process in a manner that does not keep the applicants waiting for too long. In all the Visa services I used in the past, the processes are usually well-mapped and easy to follow.

Theme 15: Risk management and legal protections

Every manufacturing and service operation has elements of risks and uncertainties. Risks management is a huge topic, but its implementation could strengthen business continuity, adaptation and resilience. Contextually, the focus here is to reflect on the perception of risks by the customer and highlight some strategies that service organisations can use to manage identified risks. Whenever risks materialise, wastes are inevitable as consequences.

When a clinical decision for a surgery was reached, a doctor called me and took me through the risks associated with the surgery and enquired if I had any questions. I replied "no". She explained that there was a need for me to sign as a consent for the surgery to proceed. I signed. Later, the anaesthesiologist approached me and explained the process of the anaesthesia and the risks. He asked if they should go ahead, and I consented. Inside the theatre, the risks were explained again even more graphically, and I still consented. The procedure was initiated. The last word I heard was when the doctor told me that the anaesthetic was about to be injected. The next time I regained consciousness was in the ward and the doctor assuredly told me that the surgery was successful. I was happy to hear that. Still on consciousness, I have imagined how I would have felt without the anaesthetic. I think that the lesson for service design is to see how unpleasant steps or activities can be hidden from the consciousness of the customer. Whereas doctors use anaesthetic to achieve no visibility, other services may use other techniques to identify and conceal uncomfortable processes from customers. Imagine that a business needs to borrow to meet the needs of a customer. Businesses do borrow, so it is not as if the business is engaging in any illegitimate activity. However, the customer does not need to know the details because it may be misconstrued as low liquidity which may degrade the confidence of the customer.

Risk is the likelihood that a loss of a value in any form might occur. In this case, the doctor listed several side effects of the procedures. In other service industries, risks can be in different forms. However, diverse risks that customers are exposed to may not be the same risks that a service organisation are exposed to. So, as a service organisation, it is not just enough to analyse organisational risks alone. I think that service organisations should also explore customer-centric risk management approach so that the exposures of the customers can be identified and managed as an integral of the overall organisational risks. This approach appears counterintuitive because corporate affairs should focus on the business and not on the customer. Interestingly, statements such as the customer being the king remains basic because all businesses exist to serve customers. To be clear about the proposed customer centric (or customer-centered) approach to risk management, it is about asking questions on all the factors that can cause a business to fail to satisfy the needs of the customer. Although the ultimate risks classification will be the same, the perception and interpretation of the organisation may differ slightly.

For instance, if the question is "what might cause us not to meet the needs of the customers?". This can still be classified as operational, supply chain, financial, environmental, IT, social, *et cetera*. However, the difference between corporation-centred and customer-centred approach is that the value stream destruction map would show the points at which values flowing to and from the customers would be at risk. Thus, the appropriate risk management approach can be deployed to smoothen the value streams.

The interesting thing about service organisations is that the survival of the client may be tied to the business continuity. During the COVID-19 pandemic, there was a Care home that, sadly, lost all the residents and the service was closed. There was another Care home that the management applied strict access to the residence, implemented COVID-19 prevention and control guidelines; and none of the residents was lost. This resulted to the service that managed the risks to protect the residents to continue to receive funding from the government while the other one was closed; the funding was terminated and staff lost their jobs.

To elucidate further the importance of customer centricity in risk management, some banks support businesses that they offer credits or loans to succeed so that those businesses can repay the loans. This is not necessarily interfering with the business of the creditor but seeking ways to support the business to be sustainable. If credit risks materialises, the bank would ultimately write-off the amount and this means that the profit would be debited to expense the bad debts. Any collaboration that will prevent a bad debt scenario would be a win-win outcome for the parties.

Nonetheless, it is not in all circumstances that a service can protect the customer from risks due to the inherent nature of risks and uncertainties.

When I wanted to travel to Prague some years ago for a conference, I bought an insurance policy to cover the trip. Due to delay in the processes, my Visa was not out on time, and I had to cancel the trip and withdraw the visa application. The risks suddenly materialised and my loses depended on the nature of the legal contracts. For the flight, it was non-refundable, and I lost the entire amount. For the accommodation, the terms required that I could cancel the booking with a 25% penalty on the total deposit. When I asked the insurance company if they could refund me the cost of the flight, they showed me a paragraph in the policy that made it impossible for them to pay under my circumstance. My losses were not due to the service design of the insurance company but the service design of the embassy that required all expenses to be made and evidence attached before visa application can be submitted. When I discussed with the embassy, they explained to me that the time available was not enough to process the Visa because they need to contact Nigerian embassy first to verify my passport. The visa fee was not refunded because the terms and conditions stipulated so. I was so unhappy that I wondered if I would wish to go to Prague in future because of my losses. Of course, I can travel to Prague if I need to, but I am trying to accentuate how customers feel when services fail and the risks affect them badly. For that trip, I dealt with different services and each of them protected itself from risks such that when the risk of cancellation of the trip materialised, I bore the bulk of the losses. On the contrary, when I applied for a Visa to travel to Canada for a conference, I thought that it was a joke when my Visa was granted and delivered to my house within two days. It was such a delightful experience!

Since service design deals with some kind of value creation and delivery, there is a risk that the value created by the service may not be enjoyed by the customer for some reasons. As of it, a service organisation must carry out extensive risk assessments of the processes for all the categories of customers they seek to serve and capture terms and conditions that would protect the organisation when risks materialise as they often do.

Risk management framework for a service organisation needs to be robust to classify risks and manage them appropriately. Risks can be transferred, absorbed, shared or avoided depending on their likelihood/probability of occurrence and consequence/impact on the service systems, environment, suppliers or customers. Regardless of the risk management approach, a service company needs to legalise and obtain approval or consent of contractual parties, individuals, groups or organisations that risks and uncertainties associated with the operations may affect.

For instance, a bank that invests customers' fund may need to be clear on whether customers are entitled to a refund of interests and principal depending on the nature of the investment. It should be made clear what the customers should expect, and the risks that are involved. An insurance company needs to be clear on what it covers and what it does not cover under certain policies. These risks and understanding should be captured as in a legal document or contract.

A transportation company that leaves the station at a fixed time needs to be clear on what happens when a customer arrives late and misses the service. Indemnities are often used to insulate the service organisation from possible risks that may result in losses on the part of the customer. For instance, with an increase in cyberthreats, an indemnity is required in situations where electronic transfer of instructions is used since organisations do not have control over how customers use and safeguard their devices and computers. Customers need to be enlightened on risks they can help to mitigate. Most customers would be happy to share a responsibility of keeping the organisation protected by taking due care not to expose self and the organisation.

Protecting the service system through legal contracts and obtaining consents are not covers for exploitation of customers. The terms of contracts of service should be fair, reasonable, justifiable, and bearable for the customers. The service organisation should identify vulnerabilities of their customers and help them to mitigate risks that may affect them. The essence of creating a legal protection for a service is to mitigate the risks of some customers that may deliberately seek to exploit the service organisation or the loopholes in the processes.

Where customers are vulnerable adults or children, there should be safeguarding policies to ensure that the customers are protected from risks associated with some actors in the service system. In this case, the legal consent is not necessarily from the customers but from staff and volunteers. Non-disclosure agreements and contracts of employment are other legal documents applicable to a service organisation and can be administered by the human resources department. Contractors to a service organisation may also grant the service some kind of protection through performance bond, guarantees, assurance, or indemnity.

Because services deal with intangibles in most cases, it is susceptible to different types of risks including IT, operational, financial, environmental, political, security, monetary, supply chain risks, etc. Lean thinking is not to be used in isolation to address these risks. Nonetheless, before a business can be efficient, it must exist first, and risk management is appropriate for increasing the chances of business continuity and sustainability. This also explains while lean and agile thinking is presented in this book in a manner that it can be useful in a wider business management setting.

I reckon that the degree of implementation of lean methodology is not as prominent in the service industry as it is in the manufacturing industry. This is partly because assigning values to intangible components of the service systems is difficult. For instance, what is the value of risk management activities in a service organisation? The value would never be known until risks materialise. At that time, it is too late. Risk management is therefore a lean tool which should be used to prevent waste generation in a service system. Risk management can also be used as a corrective tool to remove materialised risks to achieve continuous improvement.

Theme 16: Customer Chains within a Service System

A backward systematic analysis of value streams starting from the delivery of an order to the customer to the point of initiation of the order by the customer may reveal the interconnectedness and complexities within a service system. When a system is conceptualised as collection of components that work harmoniously together to transform inputs to the desired outputs, it is a high-level abstraction, and it does not capture the atomic interactivities that constitute the notion of "transformation". Nonetheless, as a systems thinker, I believe that systems thinking is fundamental to lean thinking because of the whole-system approach to lean thinking. Without adopting a systems thinking approach, the side effect of lean thinking is suboptimality in which a component of the system or a section of a process is improved but it results in a deterioration of other components of the system or other processes. So, systems thinking approach offers an end-to-end improvement and tends to reduce suboptimalities.

Regardless of how complex a service system is, people make it work. They work harmoniously to transform the inputs into the desired outputs. The inputs they transform may come from external or internal suppliers. Within the service system, as the customer requests pass through different processes, internal customer chains are activated to add value to the request of the ultimate customer. Internal customers are more visible when looking at the second layer of abstraction which is made up of departments within a service system. An output from a department will form an input into the next department(s) and so on. There are possible decoupling points to separate the customer from the control information or to route the control information to different departments for them to add value to the request. There are also stages where control information, materials and customers are recoupled. Within a service system, depending on the design, the customer, control information or material may flow from one department to another, but everything will ultimately couple together at the exit point to represent the overall service delivered to the customer. Overall, this is a dynamic process and the robustness of the service system to manage the multidimensional interactions in the system determines the performance of the system.

As an illustration of the above concept of how value flow in a service system, when I entered the A & E on Wednesday morning, the triage nurse (A) captured basic information of my case and confirmed my identity and details. The receptionist (B) then scheduled an appointment at the Urgent Treatment section. I transited from A & E to the Urgent

Treatment section. There, a receptionist (C) queued me to be seen by a GP. When it got to my turn, a GP (D) called me. She captured greater details of my case, performed preliminary examinations and then referred me to a surgeon based on her theory of what the problem might be. I waited for a surgeon to see me. When the surgeon (E) saw me, she performed further examinations and then recommended blood analysis and CT scan. I waited for the blood samples to be collected. When it was my turn, a nurse (F) collected the blood samples and sent them off to the laboratory where at least a laboratory scientist (G) will analyse the samples. I went back to the waiting area. When it was my turn, a radiologist (H) performed the scan. I returned to the waiting area. When the results were ready, I would imagine that the doctors analysed the reports and arrived at a clinical decision. Three doctors (I) further examined me in the light of all the results and informed me about that a surgery is needed. The decision was prepared, and a doctor called me and explain the findings and the decision for a surgery as my pathway for recovery. I was informed that I will stay in the hospital. A nurse (J) gave me medication in preparation for the surgery. A support worker (K) took me the ward. A nurse (L) performed few tests in preparation for the surgery and guided me on how to prepare for the surgery. The anaesthesiologist (M) explained the process of anaesthesia. A support worker (N) moved me to a different ward and later to the theatre. The surgical team (O) performed the operation. I saw myself in the ward in the morning. Nurses (P) in the ward attended to me. Support workers (Q) provided for my food, drink and beddings. When I was ready to go home, a pharmacist (R) gave me my drugs and explained how the medication should be taken. I booked a taxi and left the hospital.

In my book "Thinking, Knowing, Doing and Being", I highlighted 127 ways one can think about things. The above description focuses on the people domain because the intent is to show how an output from a staff is an input into another task; thereby creating a sort of internal customer chains. Although, the number of people that attended to me is greater than the number specified, the people or teams identified as A to R suggests that it was quite a significant number of tasks and departments/sections that were involved. The system was so robust as different processes were handled by different people, yet their activities were focused on the patient who is the ultimate customer. I was not the only one in the hospital and not everyone that came to the hospital passed through the same route as me. So, the flow of customers is not like a manufacturing line where a product has a determined flow pathway. As of it, the variability of the problems or needs of the customers in a service system contributes to the complexity in the processes in service industries.

In service delivery, the system needs to be designed with a notion of internal customers working towards meeting the needs of the staff downstream who will add their value and send the control information and the patient downstream until the customer exits the system. In the end, the customer assesses the service delivered from an aggregate view and not based on what a single department does. So, working together to meet the needs of the customer by all departments is very important in meeting the overall performance goals. When I was leaving the hospital, I did not know who to thank. I was simply grateful to all of them because all of them added identifiable value to help me get well.

Theme 17: Queuing theory and Triage model

When I arrived at the A & E, I joined the queue and progressed until I was attended to. One of the British cultures that brings peace and social order is the institutionalisation of the queuing theory. Queuing theory is a practical application of equality of human beings. Everyone knows that it is fair and transparent that one who arrives at a service point first should be attended to first based on first in first out (FIFO) logic.

In Nigeria, queuing theory is not institutionalised and who get served at the service point may depend on the social standing of the person. The concept of "Oga" or "Madam" in Nigeria is loosely translated into "a big man" or "a big woman or wife of a big man". These "big people" are ordinarily expected, albeit wrongly, to be served before others even if they arrived last at a service point. In a cultural setting like

Nigeria, one's relation or religious or tribal affiliation may be favoured at a service point. This is a concept referred to as "knowing someone". If you know someone in developing countries such as Nigeria, you will be served faster and probably in full. Nonetheless, the concept of "knowing someone "is more endemic in public services. Private businesses appear to be using queuing theory. At times, customers organise themselves based on FIFO because it is *prima facie* fair, reasonable and logical.

As a service organisation, the queuing theory is a powerful model for sequencing customer requests or order. Even in designing a software for service delivery, the notion of queuing theory can be implied using time stamp. However, for marketing and pricing purposes, the queuing theory may be adjusted to meet some performance objectives or business strategy. For instance, in train services, there is a first class and a standard service. In flights, there are first, business and economy classes. Long time ago, I heard a joke that appear to suggest that those that use standard or economy ticket will always arrive at the same time with those that used a first class ticket; and that the extra cost is not necessary. Recently, I heard an opposite joke that argued that using a first-class service is not a waste of money. It was posited that although those in the standard or economy class will arrive at the same time with those in the first class cabin, they will "arrive at different conditions". In other words, the experiences of a first class customer within the same service system differ from the experience of a standard or economy customer. Even in e-commerce, being a prime member means that you will be served faster than standard customers, although at a premium. My wife is a prime member of an e-commerce giant, and she gets her orders delivered within 24 hours but mine can take up to 5 days. From lean thinking perspectives, the processes that create the experiences for premium customers is agile and robust whereas the processes that create services for standard customers are more tuned to be lean and normal.

Service organisations can create segregations in their system but not discriminations because all customers are valuable. Segregation should not be understood in a negative sense. Segregation of services allows customers to choose how they wish to be treated or served. Discrimination favours some customers within the same class of service, and this is not good for business or public services. By segregation, the booking of a train trip or flight will automatically allow a customer to choose their preferred queue. Segregation may also be based on the customers meeting the requirements or criteria for a service. To illustrate the notion of segregation, imagine that you have a classification algorithm that can classify all the customers in your database into groups based on certain criteria. To create value from such classification, the next phase may be to assign or create unique products and services that would meet the needs of the groups. At the end of developing those products and services, the processes for meeting the needs may differ from onset or diverge at some point in the process-time-space-people domain. As such, there must be a way to segregate or separate the route of the services to be delivered by a service system at the design phase. A typical example of segregation of services is the "first class" cabin on the train. After segregation of the services and processes, standardisation can then be used to achieve efficiency or leanness. Lean thinking can be very instrumental in achieving efficiency by removing all wastes associated with serving each group of customers.

Segregation may also be based on meeting competitive criteria. This is different from grading where performance of the customers is ranked.

Segregation determines the process pathways of different customers through the service system. For instance, students of engineering can opt for MEng route at the point of entry into the university because they met the admissions requirements while other students can pass through BEng route. The programme design is slightly different even when the discipline is the same. Even before the students get to the point of making a choice, their curriculum and activities, which will determine their experiences, would have been designed.

When I was in the bank, a branch was dedicated for Personal and Private banking (PPB). The branch was for high networth individuals (HNIs) and their prestigious treatment clearly acknowledges that they are wealthy. The point about segregation and sequencing is to show all customers that they are esteemed and valued. The needs of each group of customers will still be met accordingly. For instance, a savings account customer whose average deposit is $2000 may not be interested in PPB compared to a HNI with an average deposit of $500 million. The savings customer may simply need a reasonable interest rate and unfettered access to their deposit whenever they need it. But the HNIs may want a networking lounge where he/she can meet other HNIs and discuss opportunities that are emerging on the horizon. They may even find partners for new ventures. So, they value such opportunity because it may even increase their net worth.

There is no single right or wrong approach to sequencing in service systems, but the underlying principles should be fairness, common sense and logic. Also, the service processes may be designed to allocate resources to achieve specific objectives. For instance, after the COVID-19 pandemic, many services including embassy services, hospital services, etc recorded backlogs. Imagine a situation where backlogs build in a service system for whatever reason. Should FIFO or last in first out (LIFO) be used to clear the backlogs? Based on FIFO, the customers will be treated based on the time their request arrived the service system. With LIFO, the last request will be treated first to eliminate further accumulation of backlogs. Still, the control activities may use a combination of FIFO and LIFO to clear the backlog and then maintain a normal operating rhythm or takt time (in lean thinking parlance) after clearing the backlogs.

There are some situations where the experience of the customer will not be impacted regardless of the sequence of handing tasks within the system. Such situation can be handled systematically using the most effective approach. For illustration, supposing that an insurance company desires to digitise its processes and that it requires digitising paper-based records of existing customers. The project may be planned in a way that new customers can be migrated to the new digital platforms, whilst the old records are digitised. The digitisation can be according to the year that the customers were signed on to the policy. As of it, old customers need not know that the insurance company was scanning and uploading their data. The customers may only be informed when their information have been migrated online and they are ready to use it. This will pleasure the customers and improve their experience.

There are other approaches to segregate services based on some criteria. For instance, in the hospital where people arrive with different risk exposures, the Triage system is applied so that patients with severe and life-threatening case can be attended to first. Not quite long ago, the educational system in many countries was designed to favour males compared to females. This resulted in men passing through the educational systems in numbers much more than women. Nowadays, so much have been done in many countries to redesign and re-engineer the educational systems to be equitable to both male and females. There are a couple of nations that have been held back by culture, religion, economy, knowledge-gap, etc from creating robust educational services for all people living within their country. One may wonder how this is connected to lean thinking. It depends on the level of abstraction that a country is looked at. If a country is conceptualised as a system in which products and services are created and consumed and some are exported from the system, then lean thinking can be implemented at a national scale. This was how Japan and USA became very competitive in the manufacturing sector by adopting lean manufacturing principles.

Still on a country as a system, a country that considered males and females as human capital inputs into their economy and provide equal opportunities to them may end up with a robust, resilient, diverse and productive economy compared to a nation that designed out women from contributing to their national productivity. In my hospital experience in the UK hospital, apart from the radiologist, lead surgeon, anaesthesiologist, and another two male surgeons, all the rest of the personnels that attended to me were females. In the theatre, when I was conscious of my environment, I saw a ratio of male: female as 4:3. Imagine a country where women are not supported to pursue any career of their choice, the national productivity will be lower compared to a country where every human being is considered as valuable human capital input.

In security services, triage system is used as well for prioritisation. For instance, if two threats were reported at the same time but one will result in one person dying while the other has national security implications. If resources were only available to handle one of the two threats, the threat with national impact would be handled first as a matter of priority. This may require agility, but responses need to be resource efficient. This may sound trivial, but the intention is to highlight that resources are never enough to solve all the problems facing a service system and prioritisation is a lean tool because it focuses on creating optimum value. Lean thinking is not only used to solve problems, it is also to prevent problems or create value.

There are situations in which processes may be paused safely or reorganised to attend to new requests or incidents with higher detriment. This is also at the realm of agility because it requires the service system to be flexible and adaptable. However, the system must have been pre-designed with clear classifications that would enable the operators to take swift decisions to support an "effective agility". Effective agility implies that speed may not be progress; and automating an erroneous process will increase system inefficiency.

On the other hand, "indeterminacy", (i.e., where the operators do not know how to react to needed flexibility and responsiveness) in service design is very wasteful. It could permit a catastrophic failure to happen given that staff may not like to expose themselves under indeterminate situation because they may be blamed for losses. Lack of risks management framework and lack of lean culture contribute to service system indeterminacy because such service environment may be characterised with shallow understanding of what might go wrong and how staff should respond to them. Indeterminacy is captured in one Igbo adage which states that *"ewu mmadu otutu n'enye nri nwere ike nwuo maka agu"*. This is loosely translated into "a goat that many people feed may likely die of hunger". The reason is simple. When responsibilities are not clearly assigned in a system, the staff would be looking at one another to step forward to take up the responsibility. Unfortunately, many would not take up such responsibility because such tasks may require personal sacrifice, accountability and going extra mile. In a lean cultural setting, everyone is empowered to influence the processes and they are more likely to act rationally even in novel situations. This is at the core of Kaizen principles.

The low hanging fruit for addressing indeterminacy is by creating standard procedures and protocols to improve efficiency of the service system. Let people know what to do and how what they do relate with what others do within and outside the service system. Elton Mayo also proposed teamwork and recognition as means of motivating staff to contribute and collaborate with others to create values. Even with increasing automation of processes, it has not eliminated the need for staff to work together to serve the customers. So, working together as a team will reduce the waiting time and long queues in service systems. Reduction of waiting time of the customer may result in a better experience for the customer and improved productivity of the service system.

Lastly, the queuing theory can be modified for different reasons and to achieve different objectives. From a business perspective, if a service organisation can add prestige and innovation to the same service and charge more, it makes a business sense because one of the primary goals of businesses is to make profit. In service industry, managers should treat all customers with respect, dignity, and value.

However, it is important to note that all customers are not equal in terms the value they add to the bottom-line of the business. But this should not be misconstrued to mean that some classes of customers are less valuable. Moreover, instead of considering the value added by individual customers, the value added by a class of customers should be the focus. The low-income customers may outnumber the premium customers and the greater income for a service may still come from the low-income end that requires lean thinking to serve. For instance, based on the example above. There may be thousands of customers that may have $2000 in a bank and less than hundreds with about $500 million in personal account in developing countries. Thus, the incomes that can be generated from the low-income customers may even be more than the income generated from the HNIs. So, both are equally important and the appropriate service design to meet their needs should be deployed.

Where there is a need to segregate customers to achieve profitability in for-profit service organisation, there should be no hesitation to design, redesign, configure and optimise the processes to meet the performance objectives. An average customer will be happy if the sequencing of their order follows a logical and fair processes. For quality service delivery in public services, it is a more delicate situation because there are many situations where apparent segregation of public services is inappropriate. Thus, public service delivery, which is not profit-driven may focus on standardisation of quality service delivery, continuous improvement and a transparent implementation of the queuing theory.

Theme 19: Servitisation and productisation

Servitisation is a concept that is used in manufacturing to indicate that services are bundled with physical products. It is a way through which a manufacturer can get more values from their products across the life cycle of the product. I have a 3-in-1 printer at home. When I bought it, the company offered me a service in which they will send me ink cartridge anytime it finishes but I will be responsible for the hardware and papers. They can monitor the level of ink from their office. This service has helped the company to sell their ink cartridge as well as sell more printers due to the additional services it added to the printer.

Permit me to use "productisation" to mean the concept of adding physical products to a service delivery to enrich the experience of the customer.

Although services are inherently intangible, "productisation" implies incorporating tangible products as part of the offering in a service organisation. Oftentimes, the tangible products may be offered in collaboration with partner organisations so that the service organisation can focus on their core business of service delivery. Although the service organisation may make a profit from such product, the existence of the product in a service environment enriches the experience of the customer in a convenient or cost-effective way.

Again, it depends on the service organisation and the nature of the products and how they can be integrated into the service design and processes. Also, the emphasis on products in a service environment should be minimal to avoid confusing the customers on the core business objectives of the organisation. Productisation should preferably be included based on the voices of the customers and careful research on how it can be integrated into the service delivery.

In the hospital, I do not know if the food was cooked by support staff or supplied by an external catering, but the food was a significant part of my experience in the hospital. There were shops in the hospital where patients or others can buy food, drinks, cards, etc. This is quite convenient even though things may cost higher than what is obtainable in shops outside.

Productisation of service is about getting more value from a service system by creating more values for the customer based on physical products. The customers may not necessarily pay for the products. It may be inputs into the service system, although it can be considered during costing and pricing of services. The relevance and how this can be implemented depends on the strategy, nature and scale of the service organisation.

Theme 20: Ergonomic design for services

The ergonomic design of the hospital bed was outstanding! With the console, the patient can adjust the bed for comfort and to relieve pains. The bed is an engineering product in a service environment that meets the definition of a fixed asset rather than productisation. It was carefully designed with the users in mind and it typifies the notion of customer centricity in a service environment. Engineering discipline has always provided the innovations that drive and support service delivery. Service industries should indeed continually scan the innovation horizon to spot engineering products that can enable them to delight their customers.

Think of a gym service, there are all sorts of equipment for exercise that can easily be adjusted to suit the anthropometric characteristics of different users. Such adaptable use of fixed assets in a service environment is a deliberate outcome of product design and engineering. As an example, a collaboration with researchers and companies that create ergonomic exercise kits that really work for the users will add value to the gym services. Currently, integrating sensors that can capture data during exercise and analyse them to indicate the health status of the user has been made possible through product design and engineering. The predictive nature of such products means that users can respond faster to their health and well-being needs.

Computers have transformed all sectors of service industries. However, the positioning of computers in a service environment could cause health risks due to the posture of the workers. So, the design of these tools and pieces of equipment for service delivery should consider the users and their comfort during use. This is the primary preoccupation of ergonomics and human factors in industrial and service systems.

Ergonomic designs may target staff, customers, or other users. The idea is to create lean designs that can support services whilst providing some kind of enjoyment, comfort or other values for the user due to its ergonomic design. In lean thinking, motion is a waste and it focuses on wastes due to unnecessary motion by the staff. Ergonomics may be useful in designing workstations and tools that can enable staff to offer quality services. In addition to an excellent ergonomic design of the workspace, health and safety awareness for staff on postures and other ergonomic-related stresses and ailments can be facilitated by the health and safety team. Again, ensuring that staff wellbeing is at the state that it can enable them to contribute is part of lean thinking, particularly under utilisation of talents of the staff.

Linking ergonomics to the principle of segregation of customers for the purpose of business strategy, ergonomic designs of products used for serving different customers vary. For instance, the ergonomic design of seats and spaces in a first class cabin of a train and aeroplane is different from the ergonomic design for standard customers. The material cost and manufacturing process technologies of products with high ergonomic value may be higher compared to standard product. These inputs justify why the costing for first class services within the same service system is priced differently. There are no two ways of measuring a first class service other than from the experience of the customer.

Theme 21: Single Minute Exchange of Die in Service Delivery

When I got into the theatre, I saw the brilliant surgical team that was to carry out the surgery. The team leader introduced himself and other team members. I looked around to admire the pieces of the equipment that I could see in the theatre including the lighting and layout design. Few minutes later, I was informed that the procedure was about to begin. A doctor asked me to lie flat on the bed. I was still interested in spotting lean principles. I was thinking that I would be operated on the bed that I came in from the ward. Not so, forgive my naivety because that was my first time in a theatre. When it was time to transfer to a very slim theatre bed, I experienced a lean principle – single minute exchange of die (SMED). I observed that the level of the theatre platform and the bed can be adjusted quickly to be at the same level. This made the transfer seamless, and I can imagine that, under emergency, patients can be transferred from the bed to the theatre bed in matter of seconds or in a single minute. Of course, the next time I woke up was in the bed, in the ward.

In lean methodology, changeover time matters because the time it takes to replace a tooling or start the manufacturing of a new product or switch to a new process or equipment are considered as wastes that should be eliminated. Designs and procedures can be used to reduce the changeover time to a single digit (1 to 9 minutes). Before this concept was applied in manufacturing, changeover of die or tooling could take hours to affect. Thankfully, lean thinking made it possible for manufacturers to think of how tools and processes can be configured to facilitate changeovers within a single digit minute. The SMED was developed by Shingeo Shingo and it was valuable in the Toyota production system. The scientific measurement of work including work standards, standardised work and time study were the contributions of Frederick Winslow Taylor to lean thinking. These techniques can help to standardise a process, eliminate wastes and improve productivity.

My case was not an emergency but there may be situations where ambulance service may arrive with a victim that needs to be transferred to a bed or a theatre platform for quick interventions. Time may be of essence and the application of SMED can smoothen the processes of transfers.

In logistics sector, I have seen designs where a platform can be lowered or raised during loading and offloading of goods. Although I do not know the history of the design, I observe that the adjustable platform not only reduces the loading and offloading durations, but it also offers better health and safety outcomes.

Theme 22: Design thinking in service delivery

I did not set out to write this book when I left my house to the hospital. The thought of this book came to me while I was on the hospital bed, trying to recuperate. I have discussed the impacts of waiting as an opportunity to create value. As a lean thinker, I utilise waiting time to create a kind of value where possible. Personally, I do not allow time to pass by while I watch it waste away. Perhaps this principle played a significant role in my productivity during my PhD in Mechanical Engineering at the University of Manchester, UK.

During my PhD, I created a handbook based on my research and I used it to teach over 100 secondary school students across 8 schools in Greater Manchester. I usually travel to Leeds and Sheffield for events or trainings related to the programme. I also travel to my placement locations that may take 30 minutes to 1 hour by bus for the tutorial. That is about 1 hr or 2 hours in total. This meant that I spent several hours on buses and trains. It was not long, after travelling a couple of times, that I spotted those moments that I seat in a bus looking around as wastes. So, I designed my research strategy so that I can work on an aspect of my research while waiting to be transited in buses and trains. In most cases, I may be proofreading a new manuscript, creating and refining new concepts or ideas, drafting responses to reviewers' feedback or reading a research article. When I get to the office, I would integrate the work done on transit with the workpiece in the office and then continue with it. When people hear that I published 10 articles from my PhD research in international journals, they cannot understand how that was possible. I won two awards at The University of Manchester as the MACE Department's "PGR Student of the Year" and the "Best outputs". The overall notion of my book "Thinking, Knowing, Doing and Being" is to demonstrate that it is possible to embody our thinking. I did not separate myself from the process of my PhD. I remember the day that I was having a discussion with my supervisor, and I told him that "I am the PhD" as I was trying to assure him on my commitment to the process and to highlight that my experience and learning from the process of the PhD is very important to me because that is where I will be transformed. I was not afraid of failing, I got used to some brutal criticisms by Reviewers, I often close late and resume early. I was indeed transformed by the process because it was full of learning experiences and opportunities to serve others.

I practice lean thinking and I know that it is beneficial. One of my objectives when I am teaching lean thinking is to get students to think of lean as a way of improving personal efficacy and productivity. Lean thinking, based on Bloom taxonomy is at the create, evaluate, analyse, synthesise level of cognitive domain and it could be beneficial for individuals. An effective thinker is more likely to improve manufacturing and service systems compared to an ineffective thinker. So, it will always come in full cycle that how we think influences what we know and what we do and who we will ultimately become. Notably, everything starts from thinking.

Design thinking focuses on how the design attributes of a product can meet the needs of the user. So, the design process empathises with the user so that the design process can meet the needs of the customer. The hospital bed, mentioned earlier under Theme 20, was a masterpiece of ergonomic design but its realisation was probably from a design thinking process. In order words, ergonomics is one of the considerations in a design process. With the console of the bed, the patient can adjust the height and angles of inclinations. This is not the way beds at home are designed. The persona in the design process is a sick person who may not be able to move on the bed. By factoring in all possible scenarios of what a sick person may need from a bed, such bed can be realised. Fundamentally, the adjustability of the bed is crucial in relieving pains due to posture and position in different parts of the body when moving oneself is difficult.

Apart from the bed, the electronic device for the patients' entertainment was a brilliant testimonial of design thinking. The device could provide information, knowledge and entertainment. The multi-jointed structure holding the device was adjustable to facilitate adjustment in different xyz directions by the customer. When I explored the content of the device, it has TV stations, radio stations, audio book, Internet, FAQs about the hospital and its services, directions, information on other services in the city, to mention a few.

I noticed that the audio books and movies were not free because it appeared to have been provided by a third party under the productisation concept discussed under Theme 19. However, patients that are interested in those products can pay to gain access. The radio was free, and I enjoyed the stations; although I used my smartphone to meet my other entertainment, communication and information needs.

Even though design thinking is used for products design, it can be applied in services. It can also be used to design products used in a service environment. Services and products meet need, and the design thinking approach is an interesting approach which services can adopt to design a customer centric service system.

Theme 23: Digital transformation and automated workflows

Imagine that the hospital was based on manual handling of the processes I described under Theme 16, which involved customer chains within a service. That is, the nurses and doctors will need to carry physical files of patents from one point to the other and then work back to their office or people will be employed to be moving the documents through the system so that the professionals can concentrate.

Before the advent of computers, operations used to be based on manual processes. Manual processes are still prevalent in the public sector of many developing countries. I remember when I applied for a leave as a university staff in Nigeria. I was tracking the document from one office to another. Suddenly, the paper disappeared, and it was nowhere to be found. Usually, I duplicate my application. I reintroduced another copy through the process. When I was in the bank, there was a leave portal where you can apply and track your leave application. What baffles me is the level of inefficiencies in some university systems where plethora of knowledge that change other sectors are supposed to be created. My thinking is that inefficiencies can be designed into a system or allowed to persist in a system. For instance, in some universities in Nigeria, getting a transcript of your degree result is like embarking on a new degree. The process in some schools is so inefficient and frustrating that many students have lost jobs or scholarship opportunities. In contrast, when I completed my Masters in the UK, the original copy of my transcript was attached to my original degree certificate. In the case of Nigerian universities, each application for transcript requires a repetitive process of applying, paying and waiting for months. A student or alumni centric university in Nigeria can smoothen the process of providing transcript to their students to improve their competitiveness because students often compete for opportunities with their peers from other national universities and with those across the globe. This also poses a question on how a service system might support its products to compete in the market. It may involve providing additional ancillary supports that can improve their chances of success.

In the hospital, the digitalisation of the processes is worldclass. It was mind-boggling to imagine how chaotic the hospital environment would have been if everything was based on manual handling. Digitalisation and digitisation of processes have added enormous value to the service industry. To be realistic, big corporations across the globe including those in developing nations or public sector in developed countries are digitalised. Where inefficiencies in services due to lack of IT infrastructures are still prevalent is in MSMEs across the globe and public sector of developing countries. However, with software as a service model and the revolution in software engineering, small businesses are gaining access to digital platforms to facilitate business process re-engineering. For public sectors in developing nations, the issue of corruption and the inherent bureaucratic nature of governance may slow down digital transformation of public services. When I look at some service processes in developing countries, I feel that they are implementing Poka Yoke in reverse order. Whereas the forward order requires mistakes proofing, some public sector systems in developing country are designed to be mistake prone and unaccountable.

I joined banking when it has been reasonably transformed digitally. However, during my undergraduate, I had a savings account with a microfinance bank where the operations were completely manual. One thing that was always present in the banking hall was a long queue, with people sweating and fanning themselves with exercise books. There was no opportunity for the use of mobile App or ATM card for transactions. Nowadays, it has been over 3 years that I walked into a banking hall for banking services. I can transact with mobile app, online banking, ATM card, contactless mobile payment.

Although the manufacturing sector continue to be transformed digitally, services are being transformed as well. Automation of workflows in service sector makes it easier and cheaper to attend to customers. Digital transformation enables self-service. With automation of service processes, the labour cost and errors associated with human-delivered services may reduce. Digitalisation and automated workflow do not mean rendering some people jobless. Job reduction has always been the fears of people regarding the application of automation in manufacturing and services. In fact, these fears are becoming palpable with the increasing capability of artificial intelligence (AI) tools. The emerging capabilities of AI tools will likely increase productivity of staff in the service sector. However, AI tools are merely tools and can only add value depending on how they are integrated into the service design and processes. AI cannot fully replace human beings, but they can be useful tools to improve productivity and improvements. AI tools should be used responsibly as it may reduce creativity and innovation of staff.

There is a hotel that I stay in whenever I travel to Loughborough. Once the number of days that I wish to stay is determined, I will book and pay for the accommodation. The online form is designed to show rooms that are available and their sizes. Once decision is made and payment is completed, terms and conditions, booking reference and receipt will be sent to my email. I imagine that someone will clean the room and provide the beddings, towel, soap, tea and coffee, etc. On the day of my arrival, I simply go to the reception and pick the key to the room which has been prepared for me. I have an access key that enables me to go out and come in based on my schedule. On the day of my departure, I drop the key at the reception and leave. This may likely to be like the normal procedure of many hotels. Booking and payments can be done online and in advance. This helps the hotel to plan inventory and consumables for the customers and eliminate inventory wastes.

Digitalisation and digitisation enable a service organisation to reduce human contacts which is often a source of quality defects in service delivery. In the case of the hotel, I make contact with the reception twice unless I have other problems.

In banking, digital transformation means that I do not go to the banking hall. Some service organisations provide multiple channels to enable their customers to pay for services. Customers can pay online or pay with point of sale (POS) device or pay with cash. Many non-governmental organisations (NGOs) may lack funding to fund their activities because the process of dues and donation is not digitalised. Even in public sector, taxation and levy collection systems that are digitalised may be more effective.

My children are registered with the local library, and I accompany them from time to time to borrow and return books. The greatest motivation to continue to use the library is the digitalisation of the processes. It uses self-service and you can borrow and return books on your own. I remember a couple of times that I was busy, and I could not return the books. I simply logged on to portal and renewed the books and returned them on a later date. When it remains a couple of days for the borrowing to expire, email alerts will be sent as a reminder. This is a public service, but I am absolutely satisfied with the services, and it is meeting my family needs.

The processes for services need to be systematically mapped to establish how critical elements of the processes can be automated or digitalised. Of course, there are situations that automation may not be economical. Nonetheless, service value stream mapping can reveal where manual handling in the service system is generating waste or destroying values. Such processes can be improved or automated if appropriate. There are opportunities to work with third party organisations to deliver quality service to the customers.

Theme 24: Final Words

When the inspiration for this book came to me at the hospital, I thought that it would make an interesting read. However, I did not imagine that it would take the dimension it finally took. I hope that you were able to follow the three key voices in this book which were described in the introduction session. The first voice was a narrator of my experience in the hospital and other services that I have used in the past. The second voice was an academic voice which tends to discuss, theorise, analyse, propose and criticise. The third voice is my voice of reflection as a teacher of lean thinking and my reflection on my learning journey as a student.

Allowing activities within a service organisation to evolve without a fundamental lean and agile design means that improvements and robustness will be intermittent based on the initiatives of the leaders.

The system will be effective with an efficient leader and dwindle with an inefficient leader. However, if the culture of continuous improvements, which is the corner stone of lean thinking, is strategically implemented, the improvements may be incremental, but it will lead to high efficiency and sustainable business model in the long-term instead of configuring the system for short-term profitability.

My final words are that service organisations should see lean methodology as a competition strategy and a means of transforming the culture of the workplace for sustainable corporate existence. When every staff is empowered to identify wastes and help in eliminating them, the service organisation will create more value for the stakeholders including better experiences for the customer. When customers experience the efforts to serve them better whether by introducing innovative technologies or business process reengineering or a touch of excellence on the environment, a significant number of the customers will appreciate it and may likely remain loyal. Afterall, what else do customers need other than their needs being met competitively.

Although service organisations may be in the same sector, their corporate strategy may differ. However, regardless of their strategy, there is a place for lean thinking in realising the strategic goals and objectives. Lean thinking is about causing value to flow from suppliers through the operations to the customers. Lean thinking is a systematic way of identifying, creating, transforming, preserving, transporting, delivering and sustaining value inherent in a manufacturing or a service system. Lean thinking therefore is a mixture of arts and science because values may be tangible or intangible and customers may value different things. Therefore, designing and operating a lean service system ensures that the inputs into the system are transformed efficiently to deliver values that meet the performance objectives.

So, every manager should ideally aim to configure their service design and processes to align with the performance objectives which are usually set at the strategic level of management. Interestingly, lean thinking provides a perfect link between strategic and operational levels in service organisations.

Finally, as a disclaimer, readers should not take or misconstrue any aspect or content of this book as a substitute for professional advice, a prescription or recommendation that would warrant them not to consult their professionals and consultants where necessary or conduct further research to understand their peculiar business contexts. My experiences are unique to me and may not be the same with every other person. Thus, the need for readers to consult their professionals. I do sincerely hope that you have enjoyed my interwoven experiences with several service organisations and that I have inspired you to consider lean thinking as a tool for improving personal efficacy. Most importantly, you have seen lean thinking as tool for improving service systems. I look forward to hearing your stories of what you achieved with the lean philosophy.

THE END.